The Gentrification of Nightlife and the Right to the City

Routledge Advances in Geography

1 **The Other Global City**
Edited by Shail Mayaram

2 **Branding Cities**
Cosmopolitanism, Parochialism, and Social Change
Edited by Stephanie Hemelryk Donald, Eleonore Kofman and Catherine Kevin

3 **Transforming Urban Waterfronts**
Fixity and Flow
Edited by Gene Desfor, Jennefer Laidley, Dirk Schubert and Quentin Stevens

4 **Spatial Regulation in New York City**
From Urban Renewal to Zero Tolerance
Themis Chronopoulos

5 **Selling Ethnic Neighborhoods**
The Rise of Neighborhoods as Places of Leisure and Consumption
Edited by Volkan Aytar and Jan Rath

6 **The Gentrification of Nightlife and the Right to the City**
Regulating Spaces of Social Dancing in New York
Laam Hae

The Gentrification of Nightlife and the Right to the City
Regulating Spaces of Social Dancing in New York

Laam Hae

First published 2012
by Routledge
711 Third Avenue, New York, NY 10017

Simultaneously published in the UK
by Routledge
2 Park Square, Milton Park, Abingdon, Oxfordshire OX14 4RN

First issued in paperback 2014

Routledge is an imprint of the Taylor and Francis Group, an informa company

Library of Congress Cataloging-in-Publication Data

Hae, Laam.
The gentrification of nightlife and the right to the city : regulating spaces of social dancing in New York / Laam Hae.
p. cm. — (Routledge advances in geography ; 6)
Includes bibliographical references and index.
1. Dance—Law and legislation—New York (State)—New York 2. Dance halls—Law and legislation—New York (State)—New York 3. Licenses—New York (State)—New York. I. Title.
KFX2024.H34 2012
306.4'846—dc23
2011044502

ISBN13: 978-0-415-89035-9 (hbk)
ISBN13: 978-0-415-75458-3 (pbk)

Typeset in Sabon
by IBT Global.

Contents

Figures and Tables

FIGURES

TABLES

Common Acronyms

CBs	Community Boards
CPC	City Planning Commission of the City of New York
DCA	Dept. of Consumer Affairs of the City of New York
DCP	Dept. of City Planning of the City of New York
DEP	Dept. of Environmental Protection of the City of New York
DLF	Dance Liberation Front
DOB	Deptartment of Buildings of the City of New York
FDNY	The Fire Department of City of New York
LDNYC	Legalize Dancing in NYC
LMM	The Lower Manhattan Mixed Use Districts
MARCH	Multi-Agency Response to Community Hotspots
NYCA	New York Cabaret Association
NYNA	New York Nightlife Association
NYPD	The Police Department of City of New York
SLA	State Liquor Authority
ULURP	Uniform Land Use Review Procedures

Acknowledgments

My first and foremost thanks goes to Donald Mitchell who, from the time that I began my project (which began as my Ph.D. dissertation project), has generously read my manuscripts numerous times and offered me invaluable feedback on my arguments and data collection. Even after I left Syracuse University, where I was his student and advisee, Don has never turned down his mentorship over my work whenever I requested it of him. I would not have taken the kind of academic interests or developed the kind of political consciousness that I have since I came to North America if it had not been for him. I am, therefore, greatly indebted to him. I also thank all of the faculty members at Syracuse University who participated in advising my dissertation. They are Beverly Mullings, Tod Rutherford, Authur Paris, Jamie Winders and Jackie Orr. Their feedback on my research always challenged me to refine my arguments, and became the foundation on which I have revised and developed my dissertation into this book.

This work was obviously not possible without people who, during my fieldwork in New York City, generously shared with me their experiences and the various sets of knowledge that they have established regarding the city's nightlife and the cabaret law. These people are: Robert Bookman, Paul Chevigny, David Cho, Robert Elmes, Andy Gensler, Alan Gerson, Gamal Hannessey, Kai Fikentscher, Danny Krivit, Frank Owen, Chris Rado, Tricia Romano, Shawn Schwartz, Adam Shore, Richard Vasquez, Adam X and others who have requested anonymity. I extend my thanks to the staff at the City Archive, the City Planning Commission and the Department of Consumer Affairs who kindly helped me to navigate through complicated sets of archives to find relevant data for my research. I thank Cyril Ghosh and Kris Meen, who patiently proof-read my dissertation and book manuscript numerous times, and provided insightful recommendations about how to revise contradictory statements whenever they popped up in the manuscript. Many thanks also to Yerin Kang of the Society of Architecture and Carolyn King from Geography at York University, who helped me to create and refine New York City zoning maps that appear at various points of this book. I also want to express thanks to Jason Hackworth, Scott Kidder, Everita Silina, Benedict Wallis, Pion Ltd (London), Thirteen/WNET

New York, the New York City government and the New York Nightlife Association (NYNA) for allowing me to use photos and maps for which they hold copyright. I thank Max Novick and Jennifer E. Morrow at Routledge for their incisive inputs in the matters of copyright permissions, and also for their editorial work. I also thank Michael R. Watters at Integrated Book Technology, Inc. for working on the copyedits of the manuscript. My gratitude also goes to two anonymous reviewers of the book manuscript who were supportive of this book project at its initial stage, and offered useful comments on how to improve it. I also thank the Political Science Department at York University for allowing me to work with excellent Graduate Assistants, Guio Jacinto, Laila Jazar and Muna Kadri.

Three pieces of my writing that were published previously to this book have been integrated into its content. The publishers of those works gave me permissions to do so. Chapter 8 is a revision of Hae, Laam. 2011. "Rights to Spaces for Social Dancing in New York City: A Question of Urban Rights." *Urban Geography* 32(1): 129–142, republished with the permission of Bellwether Publishing, Ltd. Chapter 7 is a revision of Hae, Laam. 2011, "Gentrification and Politicization of Nightlife in New York City," *Acme: An International E-Journal for Critical Geographies*, 10(3): 564–584, copyright held by the author with the endorsement of ACME, www.acme-journal.org. Various parts of the book draw upon Hae, Laam. 2011. "Dilemmas of the Nightlife Fix: Post-industrialization and the Gentrification of Nightlife in New York City." *Urban Studies* 48(15): 3444–3460, copyright Sage Publications.

This book is also profoundly indebted to colleagues *cum* friends in Toronto who have unsparingly offered me intelligent commentary on my research with tremendously supportive collegiality. These include: Anna Agathangelou, Greg Albo, Roger Keil, David McNally and Karen Murray. I want to express my deepest gratitude to Reecia Orzeck, Nicola Short and Jesook Song for having offered not only precious intellectual communion, but also unconditional friendship to one who has too often taken them for granted, and has never been able to pay them back. Eunhee Kim, Jihyun Lee, Moonsun Lee and Youngsoon Noh from South Korea, and Matt Crane, Eddie Robinson and Cassandra Weston from New York constantly sent me emotional support, without which I may not have finished this book. I am equally thankful for the unconditional support I have received from my mom, dad and brother (and brother's family) in Korea, who have been there for me at every step of my life.

This book is duly dedicated to all the people mentioned above as well as other anonymous ones that have enabled the completion of this book in numerous direct or indirect ways.

Introduction

Frank, a graduate student at Syracuse University, had a strange anecdote about an event that took place in a bar in Manhattan, New York City (NYC; interview, March 3, 2006). In the summer of 2005, Frank had been hanging out in an Upper East Side bar, *Mustangs*, with a couple of his friends. The throbbing music inspired Frank and his friends to swing their bodies to the rhythm of the music, until they were stopped by a bouncer. To Frank and his bewildered group of friends, the bouncer pointed to a "No Dancing" sign hanging on the wall. When Frank and his friends protested the bouncer threw them out of the bar. This seemingly unusual incident over dancing is actually perfectly legitimate, and is one that has become routine within the nightlife of NYC. This is because businesses that allow even a few patrons to dance together are subject to being fined and padlocked by the Multi-Agency Response to Community Hotspots (MARCH), a.k.a. dance police, if they are found without cabaret licenses.

Nightlife businesses that can allow social dancing among their patrons while serving food and/or drink are called "cabarets" in the city's zoning and licensing laws.[1] Compared to other nightlife businesses, such as bars, cabarets are, according to the cabaret law, subject to stricter licensing requirements for security and crowd control. At the same time, there are fewer planning zones within which cabarets can locate without having to comply with special permit requirements than other types of nightlife businesses. The stricter licensing and zoning restrictions which cabarets must submit to may appear straightforward and justifiable, given that it is logically understood that nightlife businesses that offer social dancing, such as dance clubs, cause more crowds, safety risks and noise nuisances than other nightlife businesses.

However, there has been an escalating groundswell of controversies against these restrictions over cabarets. Zones where cabarets can be legitimately located have increasingly been reduced due to the expansion of gentrification. In addition, under the Giuliani administration, first elected to power in the mid 1990s, enforcement agencies rendered it illegal for more than three patrons to "move together rhythmically" in a nightlife business that was not licensed as a cabaret, based on a strict interpretation of the definition of social dancing. That is, any form and size of social dancing made any businesses that featured or even allowed it vulnerable to heavier zoning as well as other regulations. This manner of cabaret law enforcement has cast a wide net, so that any average bar that is not set up for dancing, like Mustangs, or live rock music club where patrons often move to the music, become vulnerable to violation of the cabaret law the moment their customers start moving to a beat.

What has happened to the city that was once known to never sleep? Has NYC traveled back to the time and space of *Footloose*? Some commentators have interpreted this crackdown on social dancing establishments as a continuation of the Enlightenment project of governing through the containment of "undisciplined" bodily movements—bodily movements performed for pleasure, that are taken as a sign of being at odds with the virtues of the "rational mind" (Chevigny 2004; Ehrenreich 2007a). However, in this book I show that at the center of the controversies and conflicts over the cabaret law regulations lies a transformation of the economic and social geography of the city. As the city has experienced gentrification throughout the last three decades, "noisy" and "boisterous" nightlife businesses in gentrifying neighborhoods, including bars and lounges as well as dance clubs, have been censured as the number one enemy of "quality of life" in these neighborhoods due to their nuisance effects. Ironically, this process has gone on even as the real estate sector trumpeted and marketed the profile of nightlife in these communities as a sign of neighborhood vibrancy in order to boost property values. That is, nightlife establishments and their cultural elements have been one of the important catalysts for the gentrification of the very neighborhoods in which the presence of these businesses, later, have been intensely contested by groups of gentry that have moved here.

The rising outcry among residential communities against nightlife businesses has led the municipality to intensify crackdowns on these businesses by employing a meticulous interpretation of legal clauses applied to these nightlife businesses, such as the interpretation of what constitutes social dancing as enlisted in the provisions of the cabaret law, and proceeded with regulation in a sweeping manner. The intensifying anti-nightlife cry made the operation of nightlife businesses more and more difficult in the city. In addition, real estate prices that have risen with expanding gentrification have added more difficulties to running nightlife businesses. For sure, the difficulties have been borne across the board, even by well-financed businesses such as mega dance clubs and upscale lounges, but the repercussions of these anti-nightlife milieus have been uneven across different types of nightlife businesses. Upscale, high-finance and corporate nightlife, armed with strong financial and political capital, has survived better than the under-financed (and often alternative and underground) nightlife.

What does this changing geography of social dancing and nightlife imply about the transformation of the normative ideals of urban life? The commissioner of the Department of Consumer Affairs (DCA) of NYC once said, "[D]ancing is not regulated [under the cabaret law regulations]. The *places* that allow dancing are regulated" (Romano 2002). From this state official's optic, regulating the geography of social dancing and nightlife is not necessarily tantamount to the suppression of expressive activity, nor, by implication, the subcultural vibes in the city. However, evidence has been marshaled in the last three decades demonstrating that the decline of the geography of social dancing and nightlife means that the city has lost a site of its foundational vitality, a site of creativity, diverse expressive cultures and counter-cultural transgression of established societal norms, which in the past have placed NYC at the forefront of new ideas, cultures and radicalism. In addition, the ways in which the municipal government has administered spaces of nightlife has entailed serious violations of fundamental rights such as people's right to their sovereign bodies, alternative social interactions through social dancing, and free association to dance together and spontaneously develop subcultural communities. This further signifies that people's right to democratically appropriate and produce urban space for expression, play and socialization has been outright dismissed by the city authorities.

This book is an effort to understand this urban phenomenon, that is made plain in the struggles over nightlife and the laws and institutions that have been deployed to govern it. I investigate these struggles in relation to the gentrification, post-industrialization and neoliberalization that have continued to reshape the economic, political, social, cultural and legal geography of NYC. In particular, the book attends to how these struggles have re-configured popular perceptions of what constitute legitimate expressive activities, and re-defined the parameters of "rights" that people in cities are entitled to. That is, the story of this book tells us how recent processes of gentrification, post-industrialization and neoliberalization in (Western) cities, including those of NYC, have involved not only the transformation of urban space, but also the transmutation of values and norms that define the vibrancy and the transformative potential of urban life.

TRANSFORMATION OF CONTEMPORARY CITIES AND THE RIGHT TO THE CITY

For the last two centuries, the world over has witnessed dramatic urbanization that has taken place alongside industrialization, which has accompanied profound economic, political and cultural changes in our social relations (Lefebvre 2003). Cities have been at the center of "dialectical urbanism" (Merrifield 2002), in which urban life, space and political formations have been the drivers of the radical progressiveness of humanity, but have also been constantly engineered by new forms of capital accumulation, ideological mystification, new modes of surveillance and resultant new patterns of uneven developments. Contemporary urbanity is also beset with uneven development brought about by neoliberalization, post-industrialization and gentrification, which has also been constantly challenged by collective contestations of labor and other grassroots movements. This book examines an instantiation of such dialectical urbanism, as it has been unraveled in struggles over nightlife in NYC since the 1970s. Also, the primary problematic of this book parallels those raised by urbanists who have studied popular struggles over changing urban space and urban life in contemporary neoliberalizing and post-industrializing cities, and the implications of these struggles for the decline in people's "right to the city." The rich and nuanced details of local struggles over nightlife unfolding in this volume enrich and reinforce observations and arguments put forward by these urbanists, but also complicates and revises them.

Urban political economists have argued that the transition from Fordism/Keynesian welfare system to Post-Fordism/neoliberalism has structurally transformed the nature of contemporary urban politics, as I detail in Chapter 1. After going through suburbanization and de-industrialization in the post-war period, and fiscal crisis in the 1970s, municipalities have largely transitioned from "managerial" to "entrepreneurial" or "neoliberal" urban management (Harvey 1989) with the accordant imperative of post-industrializing cities' economies and social make-ups. Under the imperative of neoliberalization and post-industrialization, municipalities have competitively thrown up policies to lure in footloose private capital and corporate workers to raise their competitiveness in intensifying inter-urban competitions. These policies have taken on mostly pro-market and "corporate-welfare" characters, pandering to anti-union, anti-welfare and "small government" credos. This municipal policy platform has contributed to reinforcing the downward mobility of the working class and the racialization and feminization of poverty that has already been ushered in by neoliberal national policies and post-Fordist global regimes of accumulation. Urban space has also experienced similar kinds of uneven development. Gentrification of formerly derelict neighborhoods has become a consistent feature of the most prominent development policies in cities, as it provides the ground to lure middle-upper class professionals and investments by private capital into cities. However, it has also accompanied the nefarious effects of displacing economically disenfranchised working-class groups.

What has also been observed in this process is that culture has taken on a prominent role in neoliberal and post-industrial urban developments (Zukin 1995). Municipalities as well as the real-estate capitalist sector have actively appropriated aesthetic images of cities—images highlighting architecturally unique and historic environments, subcultural richness and downtown coolness—to appeal to potential middle- and upper-class gentries as well as tourists, and to enhance the images of cities in the eyes of the corporate sector. "Cultural regeneration" has been given policy priorities in many cities, especially in gentrification projects, with prominent urban scholars like Richard Florida (2004) and his "Creative City" thesis encouraging the cultural turn in urban policies. Even the more critical section of urban scholars has acclaimed the cultural regeneration of cities as the sign of a "critical social practice" (Caulfield 1994) claiming that such regeneration helps to restore urban space as a site of mixed-use diversity and social interaction between different walks of social groups, as opposed to suffocating homogenization that suburban lifestyle symbolized.

The legitimacy of these cultural strategies has not gone unchallenged, however. Urban scholars and activists have contended that the change of cities into sites of consumption of diverse cultures and lifestyles has been coupled with widening economic polarization and also with the decline of urban space in terms of being able to offer venues of spontaneous cultural expression, democracy and radical politics. A series of punitive forms of policing, exemplified by "Quality of Life" policing, have been designed to regulate "undesirable" populations (homeless panhandlers, for example)—who have come into situations of poverty often as the deleterious outcomes of the neoliberalization and post-industrialization of cities—and displace them from urban space. As regulation theorists have argued, punitive policing can be understood as a "mode of regulation," or an "institutional fix" (Peck and Tickell 1994), which is enacted in order to manage the contradictory and conflict-ridden nature of the neoliberal, post-industrial political economy and gentrification of urban space, and to re-inscribe the image of urban space as a habitat for middle-upper class gentries and as favorable sites for inward investment by global capital. The punitive nature of policing has also been imposed to regulate subcultures, land-uses, and even mundane, but essential, urban activities (e.g. minority youths hanging out with their peers in public space) when they are seen as being in the way of the quality of life and property rights claims of affluent gentries. The trend of privatizing public space (in the form of shopping malls, for example) in neoliberal and post-industrial cities has also ushered in the displacement of economically and socially marginalized populations from public spaces, and thereupon dismantled the ideal of public space as the sphere of democracy, dynamic interactions and deliberations between different social groups.

Nightlife businesses have been targets of punitive policing, as popular rallying cries against the nuisance effects of nightlife, such as the noise, vandalism by drunk party-goers, crowding, etc. that threaten the quality of life of gentrifying neighborhoods, have increased. However, the regulation of nightlife has involved a more complicated and contradictory process, as cities have actually also encouraged nightlife—as part of what I term the "nightlife fix" in declining urban economies. For example, the central and local governments in Britain have been actively promoting nightlife through the "24-hour city" concept to revitalize depressed urban economies and derelict downtowns. In the United States also, if to a lesser degree, diverse initiatives have been designed to promote nightlife as a pioneer of gentrification and to market nightlife as the symbol of urban vibrancy and subcultural diversity. Nightlife studies have identified that this contradictory approach to nightlife—deregulating it at the same time as re-regulating it—has ushered in the "gentrification of nightlife" in which underground/alternative nightlife venues that have developed alternative philosophies of communities and subcultures have been marginalized, and profit-driven, upscale/corporate forms of nightlife have prevailed (Chatterton and Hollands 2003; Hadfield 2006; Talbot 2006).

The disappearance of spaces for transgressive and alternative subcultures implies a serious decline of people's rights; that is, people's rights to appropriate urban space and participate in producing it for the purpose of use value, play, diverse social interactions, alternative community-building and the radical re-imagining of urban society (Hae 2011a). As Mitchell (2003) and others have argued, the militarization of urban space in general has brought about new cognitive maps in urban society about what should be the legitimate "rights" that urban inhabitants are entitled to in their life-world and life space. In the new urban conditions, in which neoliberal and post-industrial directives have prevailed, the civil liberties, civil rights, public spaces and social interactions necessary for the development of democracy are conveniently expendable. Cities that are supposedly culturally rich and diverse have become a site in which the rights of the privileged few whose property rights concerns and whose politicized claims to quality of life trump other rights central to the "normative ideals of urban life" (Young 1990) such as democratic access to multiple urban spaces (including spaces of play and for use value), democratic participation by diverse individuals and groups in the production of urban space and enjoyment of a diverse social/cultural life and the kinds of socialization that are unique to cities.

However, as Mitchell (2005) showed, the conditions that realize these normative ideals—such as protecting mundane, basic activities in urban space—are not usually captured as constitutionally protected rights under liberal legalism (i.e. these mundane activities are not easily recognized as "speech" or "expression" that would entitle them with constitutional protections), which makes it hard for activists to legally challenge the gentrification and punitive policing that threaten to unduly regulate these activities. This explains the emergence of urban movements based on the "right to the city" (Lefebvre 1996) as a radical alternative to the rights frame that undergirds liberal legalism. The "right to the city" has been used as an organizing principle to mobilize people who seek to establish an alternative society of radical democracy in which people's rights to democratically create and appropriate spaces of use value are secured against colonization by market rationality, as well as the state's undue infringement upon them. This book also shows how re-imagining within the "right to the city" framework spaces for mundane activities, such as nightlife and social dancing, can provide a crucial tool through which one can reformulate popular understandings of the importance of (spaces) for these activities (however trivial they may seem), democratic participation among diverse urban inhabitants in production and appropriation of urban space for play, and the need to question the legitimacy of the state's sweeping regulations of these valuable urban spaces. That is, the "right to the city" can provide a useful platform of popular challenges to uneven development and decline of democracy that have emerged under neoliberalization, post-industrialization and gentrification processes.

NIGHTLIFE, GENTRIFICATION AND THE RIGHT TO THE CITY IN NYC

This book examines the regulatory regime that has sought to control and manage nightlife in NYC. For this investigation, the ensuing chapters in this book detail the "legal complexes"—"the assemblage of legal practices, legal institutions, statutes, legal codes, authorities, discourses, texts, norms and forms of judgement" (Rose and Valverde 1998: 542)—related to the cabaret law, especially its zoning regulations, and other laws that have been enforced on nightlife. In these chapters I investigate how such legal complexes have been unfolded in relation to the increasing efforts on the part of the city to neoliberalize and post-industrialize the urban economy, attract the "creative class," gentrify formerly derelict urban spaces and market cities as vibrant subcultural centers with a flourishing nightlife. And, I attend to how they have been at the center of struggles between different local actors who have advanced conflicting ideas of citizens' rights as they are implicated within the context of gentrifying urban space. In particular, I offer an analysis of how particular patterns of politicization among pro-nightlife actors, a hitherto under-researched subject in nightlife studies, has been emerging in relation to gentrification and Quality of Life policing in the city, and also in relation to uneven development between different nightlife sectors. This investigation discloses how broader political economic processes—in this case, neoliberalization, post-industrialization and gentrification—are embedded into, and also constituted by, militarizing urban space, and changing mundane social relations and popular perceptions of what constitutes legitimate expressive cultures within the city's boundaries.

Following Chapter 1 where I examine how this book reflects and complicates the existing theories of changing urban conditions shaped by neoliberalization, post-industrialization and gentrification, and the "right to the city," the book flows in a largely chronological order starting from the 1970s, in order to show how the city's nightscapes and the subcultures associated with them were shaped along with the changing urban political economy and social relations characteristic of each period. In Chapter 2, "The Cabaret Law Legislation and Enforcement", I introduce the racist origin of the cabaret law that was particularly discriminatory towards live jazz music, and also the institutional matrix and administrative procedures within which amendments and enforcements of the cabaret law have been carried out in NYC. Chapter 3, "Development of Dance Subcultures in the

1970s", recounts how when the cabaret law enforcement was relaxed during the fiscal crisis of the mid 1970s, an underground dance music scene developed in abandoned quarters of the city in relation to a particular body politics among mostly (but not exclusively) blacks and gays, and became popular, later, in the mainstream as disco. In terms of the latter, I examine midtown celebrity oriented clubland, the beacon of which was Studio 54, and how it shaped popular perceptions of dance clubs in general. The chapter also reflects on disco's contribution to the post-industrialization of the city's economy, culture and space, despite controversies that it started to bring about in such gentrifying neighborhoods as SoHo and NoHo.

In Chapter 4, "Gentrification with and against Nightlife: 1979—1988", I show how nightlife induced gentrification in extensive spans of neighborhoods such as the East Village, the Flatiron District and TriBeCa, by helping to revalorize depressed properties and enhance the appeal of these neighborhoods. This is juxtaposed with an opposite story about how gentrification exerted a crippling effect on subcultural communities that were the architect of the flourishing nightlife of the city. I call the process "gentrification with and against nightlife." In conducting this analysis, I document the rising anti-disco outcry of residential communities, and subsequent tightening of the cabaret law regulations by the Koch administration as part of Quality of Life policing from 1979 to 1988. I demonstrate that small/underground/bohemian dance/music clubs suffered most by this crackdown. This chapter also elaborates on the legal reasoning articulated in a lawsuit, *Chiasson v. New York City*, brought by the Musicians' Union to challenge the constitutionality of the cabaret law, as tightened cabaret law enforcement was having a detrimental effect on live jazz music businesses.

The aftermath of this lawsuit is analyzed in Chapter 5, "Zoning Out Social Dancing: The Late 1980s". In this chapter, I detail the process of rewriting the cabaret law zoning regulations in the late 1980s and early 1990s that was initiated as a result of the *Chiasson* ruling. The revision involved a substantial loosening of the zoning regulation of live (jazz) music venues, while drastically reducing zones that allowed businesses that would have social dancing of *any* size. This chapter discusses the revision in relation to the precarious constitutional standing of social dancing *versus* live music as a constitutionally protected expression. I argue that this revision should be understood as a "mode of regulation" that the municipality employed in order to stabilize the emerging post-industrial economic regime and gentrifying landscapes by displacing the "disorder" that erupted at the time of such urban economic change. This chapter also describes the gradual decline of downtown club subcultures in the late 1980s.

In Chapter 6, "Disciplining Nightlife: 1990—2002" I show that the period identified in the title experienced an explosion of conflicts over a wide variety of nightlife businesses—bars, restaurants as well as dance/music clubs—as expanding gentrification attracted more nightlife businesses. I describe a range of laws proposed or legislated by municipal as well as state governments to cope with these conflicts. This chapter also describes how the Happy Land fire in 1990 has become a key event in the history of nightlife regulation that has shaped the official as well as popular conception of nightlife businesses that offer social dancing, and the regulatory approach that has been taken towards them, in parallel. This is followed by a discussion of the stepped-up regulation of nightlife taken by the Giuliani administration—a vigorous proponent of neoliberal economic policies, and of Quality of Life policing and the Zero Tolerance initiative—and in particular, controversies over the sweeping regulations imposed on nightlife, including one in which the enforcement agency penalized businesses based on the social dancing provisions of the cabaret law that allowed more than three people moving rhythmically together, if the business did not hold a cabaret license. This chapter further examines discourses circulated by anti-nightlife coalitions that effectively disciplined the city's nightlife into more gentrified forms and imposed "subcultural closure." I situate these discourses within the context of revanchism, popular cultural politics and tactics of vengeful management of "undesirable" populations and space that have been gaining hegemony in neoliberal and post-industrial cities.

Chapter 7, "Voices for Change: From 2002 Onwards", examines the rise of pro-nightlife activism in the city, which has not yet been rigorously studied in the field of nightlife study. I analyze tensions that arose between Legalize Dancing in New York City (LDNYC), an anti-cabaret law activist group that protested the social dancing regulation of the cabaret law, and the New York Nightlife Association (NYNA), a pro-nightlife lobbying organization that is primarily comprised of owners of mega dance clubs, upscale lounges and bars—tensions which surfaced as the municipality attempted to reform the cabaret law and initiate new nightlife license laws. This chapter shows how the tensions between these two groups had a bearing on the uneven development among different nightlife sectors. Based on an analysis of the limited nature of each group's pro-nightlife politicization, I argue here that a more robust and comprehensive challenge to gentrification itself and the wide-ranging Quality of Life policing of nightlife is needed within pro-nightlife movements.

Chapter 8, "The *Festa* Ruling, the Right of Social Dancing and the Right to the City" analyzes a lawsuit which took place in 2005, *Festa v. New York City*, filed by dancers, dance organizations and anti-cabaret law activists to challenge the constitutionality of the cabaret law's regulation of social dancing. Based on the analysis of the *Festa* reasoning, I argue that the court's "categorical approach" to First Amendment protection signifies "judicial anti-urbanism," in the sense that the approach aids the suppression by the municipality of mundane urban activities, associations and their spaces, like (spaces for) social dancing, which are invaluable to democratic urban life but are rarely recognized as a constitutionally protected form of "expression." I re-introduce, here, the concepts of "urban rights" and the "right to the city" as principles that can help us to theorize and politicize the necessity of protecting those activities, associations and their spaces, as essential conditions of constituting the normative ideal of urban life.

Following this, the "Conclusion" revisits the contradictions of the "creativity fix" by reviewing the history of NYC nightlife regulations, including the cabaret law regulations, in which people's rights to creative subcultures and diverse forms of socialization encapsulated within nightlife, and people's rights to democratically create and appropriate spaces of use-value, play, social dancing and nightlife, have been seriously compromised. I argue that it is important to take seriously the suppression and disappearance of particular urban activities and their spaces, as these are invaluable in establishing the normative ideal of cities. I conclude with the assertion that while my empirical case is concerned with social dancing and nightlife, the implication of this research can nonetheless be applied to urban movements that have fought against broader neoliberal and post-industrial offensives against liberal/social rights and the notion of democratic access to urban life and space.

METHODOLOGY AND OBJECTS OF STUDY

The geographic focus of this book is the area of Manhattan below Central Park, and its historical focus spans a period encompassing approximately the end of the 1960s to the late 2000s. I have selected this spatial and temporal range because during this period many dance clubs and other nightlife businesses have concentrated in the once-abandoned neighborhoods below Central Park, and this area has been aggressively gentrified since the late 1970s, thus leading to several intense conflicts over nightlife businesses. Despite this specific focus on neighborhoods below Central Park, the main object of scrutiny in this book remains NYC in its entirety as the laws that this book examines have been applied to all nightlife businesses in the city. I also mention in Chapter 6 how gentrification in Manhattan has pushed nightlife businesses into other boroughs, especially Brooklyn, and how nightlife businesses in other boroughs have gone through similar neighborhood conflicts when these boroughs also started to experience gentrification.

This book is based on ethnographic research that I conducted in various phases between 2002 and 2009. During this time, I collected data on the history of the cabaret law. While the cabaret law is a blanket law that involves several departments in the municipality, my data collection was primarily focused on the Department of Consumer Affairs (DCA) and the City Planning Commission (CPC). The focus on these two departments is due in part to the fact that these two are the main governmental bodies that have legislated, or amended, provisions of the cabaret law and, as such, have been embroiled in turbulent battles with civil society groups when legislation was created and enforced. In addition to the list of currently licensed cabarets in the city, I acquired from the DCA, the Administrative Code, Rules, and transcripts of public hearings. From the CPC, I acquired digitized historical zoning maps starting from 1961 and reports on the amendment of the cabaret zoning. The zoning maps were reconstructed in order to visualize how (radically) the 1990 rezoning of the cabaret law transformed the locations where dance venues would be allowed and under what conditions. The municipality's regulatory approach to problems of dance venues and other nightlife businesses was analyzed on the basis of local newspaper reports (especially the *New York Times* and *Village Voice*), CPC reports, transcripts of public hearings, municipal communiqués stored in the Department of Records (also called the City Library), and interviews with officials in the DCA and the CPC. The majority of the municipal documents stored in the Department of Records that had relevance to the cabaret law and other nightlife regulations were newspaper articles published in local papers. This is the reason why this book many times cites newspaper articles.

I also conducted interviews with public officials and civil servants in the DCA and the CPC. These officials provided me with useful information about zoning amendments and explained to me the approaches that these departments were taking with respect to the cabaret law and other nightlife regulations. Unfortunately, however, these interviews yielded limited information. Those who are currently working in the DCA and the CPC in the area of cabaret licensing and zoning are all relatively new because these personnel are frequently rotated between different departments. Most of the public officials and civil servants that I got in touch with were not well versed in the history of cabaret law legislations, and they routinely referred me to written data sources, such as newspaper articles. Second, the period of my fieldwork was a decidedly sensitive time for discussions with public officials about the cabaret law and other nightlife regulations, given that the municipality was at the center of media attention due to increasingly strong activism from pro-nightlife actors. On many occasions, my requests for interviews were refused by high-ranking governmental officials, who apparently sought to avoid conversing about sensitive issues.

The controversies and conflicts over the operation and location of dance clubs and other nightlife businesses have been first identified and analyzed based upon the information gathered from media coverage. Additionally, I was a subscriber to two internet listserves—one operated by an anti-cabaret law group, *Legalize Dancing in NYC* (LDNYC), and the other by a pro-nightlife organization composed of owners of bars and restaurants, *Taverners United For Fairness* (TUFFNYC). I was allowed to quote discussions and debates among members of these listserves as long as citations remain anonymous. I also interviewed three members of the LDNYC. I have tried to interview members of another nightlife organization, the *New York Nightlife Association* (NYNA), a trade organization for nightlife businesses, but have only been allowed to interview the lawyer who represents the NYNA. But, I was able to interview two members of the NYNA, one of whom was also a member of LDNYC. I requested interviews with members of four Community Boards (CBs) located within the boundaries of the focus area of my research, but these requests were either declined or greeted with no response. Most of the information regarding CBs, therefore, comes from media sources, and my participant observation of two public hearings where CB members were present. Transcripts of public hearings were used to identify the main participants in the neighborhood conflicts and discourses advanced by these participants. Documents related to the 2005 lawsuit that challenged the cabaret law's constitutionality were acquired from the lawyers defending both plaintiffs and defendants. I also conducted several interviews with one of the plaintiffs' lawyers. For the history of conflicts over issues of nightlife (regulations) before 1989, I relied upon newspaper articles and books, which were complemented by official documents (e.g. transcripts of public hearings). An organization that represents nightlife businesses and campaigns for pro-nightlife agendas did not emerge until 1989, and individual nightlife actors that I interviewed did not recollect the detailed contents of nightlife regulations in the 1970s and 1980s (especially over dance clubs), which consequently made the media/literature review the key resource of scrutiny for these older periods.

Regarding the history of dance clubs and other nightlife businesses in the city, the main source for my analysis was the existing literature on the subject. This analysis of the existing literature was supplemented by interviews with one ethnomusicologist, three club owners, one record-store owner, three DJs (who primarily DJ-ed house and techno music), one promoter and two *Village Voice* writers. To come to grips with the topics of the clubbing experiences of nightlife patrons (especially as mediated through social dancing) and the implications of these on the reconstitution of their individual/collective identities and senses of spatiality, temporality and sociality, I resorted primarily to the three major ethnographic studies that have been published about NYC's clubland (i.e. Buckland 2002; Fikentscher 2000; Lawrence 2003). My own research, however, did not entail research contributing to theories of the micro-scales of the clubbing experiences. There has been criticism that many nightlife studies have focused primarily on regulatory changes, and that in doing so, clubbing/drinking experiences have been represented in an overly simplified way (e.g. Eldridge and Roberts 2008; Jayne, Holloway, and Valentine 2006; Jayne, Valentine, and Holloway 2008; Shaw 2010). Admittedly, nightlife is a multifaceted phenomenon requiring serious analysis in terms of the biological, the technical, the cultural, the social, as well as the economic and the political. While I am sympathetic to theoretical and empirical insights that have explored multiple and fluid forms of spatiality, temporality, and sociality that are involved in the "experiential consumption" (Malbon 1999) of nightlife, and clubbing and drinking (e.g. Redhead 1993; Redhead et al.

1998; Thornton 1996), the research that this book is based on differs from this type of scholarship both in its objectives and in its methodologies.

In other words, the primary goal of this book is to investigate the ways in which the urban environment, restructured under neoliberalization and post-industrialization, has been reshaping the *conditions* in which the exercise of people's rights to (spaces for) "experiential consumption" (of night clubbing and social dancing) are thwarted, and corporatized/gentrified forms of nightlife become the primary provider of nightlife to people. In addition, this study analyzes how pro-nightlife formations have politicized themselves in the three-decade long struggles over NYC nightlife and social dancing, and in what ways pro-nightlife politicization has been limited in developing a more robust and comprehensive political response to the conditions threatening the city's nightlife. For this reason, this research primarily focuses on the narrative of conflicts over nightlife regulations, and conveys the voices of people who have been directly involved in the fights for, or against, the creation of such conditions—including but not limited to NYC government officers, community board members, owners of dance clubs and other nightlife businesses, lawyers, journalists, pro-nightlife activists and anti-cabaret law activists (the latter comprises cultural critics, ethnomusicologists, clubbers, performers, DJs, musicians and so on). The question of pro-nightlife politicization has been under-theorized in nightlife studies. This subject deserves serious scrutiny as studying the politicization of nightlife enables us to better calibrate the political agency and transformative potential of nightlife, and to re-situate nightlife formations as part of broader urban movements that fight against gentrification. I believe my analysis of the politicization of pro-nightlife formations would make an important contribution by pioneering a hitherto under-researched and under-theorized subject.

My research threads together fragments of findings from archival research, document analyses, interviews, and participatory observations to provide a comprehensive history of nightlife regulations in NYC. A historical materialist framework—especially as it was developed within urban geography—has been the theoretical guide throughout my ethnographic work. But following feminist historical materialist critiques of abstractionism and reductionism, I have also tried to produce a more comprehensive picture of social totality, by providing dialectical forms of explanation articulating the necessary and contingent social conditions, a rich, nuanced sense of historical and geographical contexts and the multi-faceted and over-determined nature of struggles over nightlife. The chapters that will follow, therefore, show the complicated and contradictory processes of the actually existing urban reality.

1 Transformation of Urban Space and the Right to the City

While the term "gentrification" was coined in 1964 by Ruth Glass, the process of urban regeneration, of which gentrification is one form, has existed for a long time, and its features have both mirrored and shaped broader economic and social changes of each period. Gentrification since the mid-1970s in North American and British cities have also been shaped both by the broader economic and social changes resulting from a transition from the Fordism/Keynesianism complex to the post-Fordism/neoliberalism one, and by the concurrent rescaling of politics and economy between national, regional, and urban levels. This chapter examines how scholars have discussed, and debated over, the transformation of urban space under neoliberalization and post-industrialization. In particular, I attend to how these debates and discussions have been unfolding with respect to the processes of gentrification and militarization of urban space. This chapter also examines how gentrification and the militarization of urban space have been related to gradual changes in notions of justice, citizenship and the kinds of rights that people are entitled with, as well as the governmentalization of market rationality into wider fields of urban society. Studies of nightlife, which have researched the gentrification of nightlife in particular, show how crackdowns on nightlife are related to the broader move of punitive policing that is characteristic of cities that experience neoliberalization, post-industrialization and gentrification. This book is situated within these scholarly discussions, but I also maintain that the empirical and theoretical arguments that this book makes intervene in, complicate and enrich these discussions and debates.

GENTRIFICATION IN NEOLIBERAL AND POST-INDUSTRIAL CITIES

The post-war societal order in North America and Britain was broadly based on the Fordist regime of accumulation and the Keynesian welfare regulatory system. With the economic sphere revolving around a national structure of production and consumption and generally taking after a form of planned economy, the state, the capitalist sector and organized labor formed a social contract in which the working classes were granted higher wages, full employment, welfare benefits and social services by the state and capitalists in return for high productivity (Jessop 2002).[1] By the end of the 1970s, however, an array of political, economic, financial and social change at the domestic and international scales (including the OPEC oil crisis in the early 1970s) ushered in the crisis of the post-war order. In the United States, profit rates started to decline and manufacturing industries started to internationalize their production process to low labor cost regions. The way that state governments sought to overcome the crisis and especially the inflationary spiral was by undermining the power of the working class and the welfare state philosophy, and by restoring market discipline over labor and social policies. This was the context in which neoliberalism was adopted both in Britain and in the United States as a regulatory orthodoxy that would replace Keynesian welfare statism under the slogan of "There is No Alternative" (TINA).

The political mobilization of neoliberalism was coupled with the emerging regime of flexible, or post-Fordist, accumulation that came into being out of the Fordist crisis. Post-Fordism was characterized by the casualization of labor for precarious forms of work and frequent use of outsourcing. Neoliberalism emerged as regulatory principles at the national and international scales and facilitated the emerging post-Fordist system. Regulation theorists argue that a specific capitalist regime of accumulation—which is defined by norms of production, distribution and exchange, labor organizations, and a technological paradigm—is constantly unstable, contradictory, conflict-ridden and crisis-prone, and thus, in order to secure a semblance of order, coherence and stability, regulatory management is required (Lipietz 1994). The latter is called "mode of regulation," and it refers to institutions and social systems constituted by a new set of discourses and representations that work to (re)structure people's subjectivities in order that they become compatible with the general logic of the regime of accumulation.[2] When stability cannot be maintained any more, a new regime of accumulation emerges accompanied by a new mode of regulation. The transition from Fordist to post-Fordist accumulation was also coupled with the change in the mode of regulation from the Keynesian welfare system to the neoliberal system (Peck and Tickell 1994).

With the change to the neoliberal regulatory system, the Keynesian tenet for welfare and social provisioning by the state was gradually dismantled in favor of fiscal discipline, and privatization and tax reduction became legitimate and ideologically preferred policies. The argument that employment security, "lazy" welfare recipients, "greedy" public sector unions, and the social rebellions of the New Left during the 1950s and 1960s—such as civil rights and anti-war movements and identity politics-based movements—were barriers to capital valorization and, therefore, causes of the 1970s crisis acquired a discursive hegemony (Tabb 1984). Over the last three decades of neoliberalization in the United States and in Britain in which an array of new institutions, regulations, socio-political alignments and socialization were experimented with and implemented, the working classes experienced downward mobility, a decline in full-time employment and a rise in the rate of exploitation, while a series of deregulation by states granted enormous power and free mobility to the hands of the capitalist classes, particularly finance capital. This is the context due to which David Harvey (2005: 15–19) has argued that neoliberalization has been a project of restoring class power by the capitalists and the bourgeoisie.

While regulation theorists have focused on the macro- and meso-level analyses of regional, national and supra-national institutional transformations and corresponding socio-political re-alignments, Foucauldian theories have focused on the microphysics of neoliberal ideologies at the scales of the local, communities, everyday, body, as well as the scale of state apparatuses. Governmentality refers to governing from a distance, governing through particular rationalities that are made diffused throughout an assemblage of legal, political, architectural, financial, administrative institutions and social relations, and governing through subject formation rather than repression and punishment (Brown 2003: fn. 2; Rose 1996). Neoliberal governmentality has been enabled by disseminating market rationalities—such as the prioritization of profitability, privatization, the value of utility, property rights—and an ethos of individualism to the fields of the social, the political and the moral (Brown 2003). It also seeks to produce and reward the entrepreneurial, voluntary, and responsible citizen-subject that is capable of self-care and self-management of risks. Those who do not conform to this subjectivity and represent "dependency" are rendered disposable and undeserving of legal protections (Giroux 2006).

Neoliberalism is a multi-scalar and multi-faceted process, an observation that underscores the importance of combining analyses of various theoretical trends, such as political economic, regulationist, and governmentality theories (Larner 2000). In addition, there have been voices that call for more context-sensitive understandings of neoliberalization (Ong 2007). Neoliberalism develops in constant tensions, enmeshment, and hybridization with inherited institutional legacies and existing social-politico constellations of power in particular locales, and therefore, "actually existing neoliberalism" (Brenner and Theodore 2002) is path-dependent, uneven, and variegated across different locales (Brenner et al. 2010). Neoliberalism has also evolved and adapted since its wide adoption by politicians in the 1970s, as regimes that adopted neoliberalism have constantly confronted crises after crises, and have implemented new forms of neoliberal policies as crisis-management mechanisms. Therefore, any trans-historical characterization of neoliberalism would need serious revisions. For example, Peck and Tickell (2002) argue that neoliberalism has evolved in two distinct phases: the first was characterized by "rolling back" of a range of Keynesian welfare state institutions, which was followed by the second phase from the 1990s in which "rolling out" of a series of neoliberal regulatory systems was carried out.

Cities have been prominent sites where variegated forms of neoliberalization have been experimented with and contested. Municipalities have struggled with dwindling revenues caused by deindustrialization and massive suburbanization since the 1960s. Financial constraints for these cities also grew sharply after the fiscal crisis of the mid-1970s and the subsequent neoliberal downsizing of federal outlays to cities. These changes imposed financial uncertainty on municipalities, and private capital has, therefore, emerged as a prominent financial source to which municipalities have had to turn. This shift has embroiled municipalities in fierce inter-urban competition to market themselves to attract increasingly footloose capital and corporations (Brenner and Theodore 2002). Swyngedouw (1992) has called the resulting rescaling of politics "glocalization." Municipal financial structures are evaluated by bond-rating agencies such as Moody's and Standard and Poor's (S&P). Interests of private capital review these evaluations to decide whether to make a loan to the city or otherwise invest in it. For the most part, these agencies consider strong fiscal discipline to be yardsticks of good governance and economic healthiness of municipalities and a low-risk investment option (Hackworth 2007: 17–33). This practice, together with the global ideological changes in favor of neoliberalism, has had the effect of disciplining municipalities' fiscal practices, leading to the vehement adoption by municipalities of a neoliberal and entrepreneurial approach to urban governance, and consequently seriously weakening urban autonomy. Municipalities prioritize market interests as the most prudent governing principle, and public-private partnership and privatization projects have proliferated in recent years with a concomitant retrenchment of socially inclusive public policies (Brenner and Theodore 2002; Hall and Hubbard 1998). This approach stands in contrast to the "managerial" approach (Harvey 1989) taken under the Keynesian welfare state system, in which redistribution, provisions of collective consumption and fulfillment of public interests were equally central policy platforms of municipalities.

As municipalities have sought to provide favorable business climates and "corporate welfare" in the form of infrastructural provisions, labor regulations and tax incentives, and as welfare responsibilities were downloaded from higher level to lower level governments without matching financial resources, investments in social service and welfare provisions to the working class were dramatically reduced at the municipal level, effecting the racialization and feminization of poverty (Wacquant 2001). One way that emerged for municipalities to combat poverty was through inculcating the sense of self-help, entrepreneurship, empowerment, and active citizenship by having individuals and communities in deprived areas take greater responsibility for their own well-being (Raco and Imrie 2000).

Capitalism needs a "spatial fix" to overcome its crises and secure its stability (Harvey 1982), and one of neoliberalism's paradigmatic spatial fixes at the urban scale has been real-estate (re)development, most frequently in abandoned inner-city areas and former manufacturing districts. As profit rates fell drastically for manufacturing industries in the 1960s and 1970s in Western "developed" countries, real-estate development has emerged as one of the new fields for investments in an increasingly deregulated global financial market.[3] The drive for property redevelopments also comes from the fact that municipalities have very circumscribed tax raising capabilities to meet local demands of high expenditure and to rejuvenate depressed urban economies. Apart from flat rate income taxes, property taxes are one of the major sources of revenue for municipal governments (Hill 1984: 302); therefore, campaigning for commercial area redevelopments and gentrification has become a fiscally pragmatic choice for municipalities. Municipalities have thus legislated various deregulatory laws and incentive programs to facilitate private investments in real-estate developments and minimize non-market interferences (Fainstein 1994).

Neoliberal gentrification has evolved taking on different features at different times (Gotham 2005; Hackworth and Smith 2001). It started roughly in the 1950s, but this first wave of gentrification represented only very small groups of gentrifiers. The second wave of gentrification that ensued after the 1970s recession was characterized by public-private partnership and market-centered policies, the integration of gentrification into the global financial system (cf. Weber 2002), and intense resistance to gentrification among grassroots. Since the early 1990s recession, a third wave of gentrification has been observed in which the presence of corporate developers and real-estate investment trusts (REITs) in gentrification became conspicuous, municipalities became more assertive in subsidizing gentrification, and militant resistance to gentrification faded out.

On the other hand, with the urgency to post-industrialize the economy on the part of municipalities, the priority of urban redevelopment moved away from provision of affordable and social housing for the poor and working classes. Instead, development of office buildings for postindustrial businesses has settled as the focus of these redevelopments. These businesses are typically service-oriented and information/knowledge-intensive and belong to the culture, finance, insurance, advertising and other such industries. The development of office complexes is accompanied by the development of upscale housing and consumption/leisure sites for the middle- and upper-class professionals who work in these businesses, and whom Richard Florida (2004), a renowned scholar who initiated the "creative city" fad in the field of urban public administration, has called the "creative class." It has been well-documented that low-income and working-class people have been displaced from the central cities (Newman and Wyly 2006), and the shortage of affordable housing for the working poor in the central city has ushered in a massive increase in homelessness and the number of people who are precariously housed (Mitchell 2003). The reformist campaigns for "social mixing" gentrifications proposed as a policy response to this housing problem have proven that social mixing is another name for "displacement by gentrification" of low-income populations (Lees 2008). Despite the externalities of gentrification that have continued unabated, real-estate (re)development is still one of the most paradigmatic projects for city governments, in that it attracts private capital's investment interests, brings tax income into the city, and provides the physical infrastructure for the restructuring of the city's economy in the direction of post-industrialization. There remains a strong sense among city officials and politicians that such real-estate (re)development is a medium through which cities can raise their competitiveness under conditions of rapidly intensifying inter-urban competition.

CULTURALIZATION OF GENTRIFICATION

For the last three decades, scholarly work examining various causations and socio-economic impacts of gentrification has proliferated (see Lees et al. 2010). Clearly identified and re-confirmed in this expanding literature is the fact that gentrification is a multi-dimensional process, fraught with conflict and tension, and contingent upon geographical and historical contexts of specific locales. Here, I want to discuss one example that reconfirms some of these points, i.e. how "culture" has been strategically placed in the process of gentrification and to what contradictory effects.

Under Western post-Fordist regimes, cultural differences and idiosyncratic identities have garnered distinctive sign values for commodities that would satisfy consumers fitting in niche markets (Klein 2000). This contrasts to the broadcast model of consumption that, in general, produced homogenous goods and lifestyles under Fordism. The change under post-Fordism is related to the fact that the baby boom generation grew up in the counter-culture milieu of radical identity politics as well as mass consumption, a generation which had, by the 1980s, come to constitute the majority of consumers. The social and cultural capital of these new and numerous members of the middle class in the 1980s became closely associated with tastes for differences, individuality and diversity. These tastes have all been appropriated by markets under post-Fordist regimes (Christopherson 1994: 414). These features of post-Fordism have been reflected in, and mediated through, new urban landscapes from the 1980s.

As city governments have sought to post-industrialize the urban economy, "culture" has taken a more prominent place that would supposedly resolve urban problems. Cities that have suffered from an image of dereliction and despair since the 1970s started to aggressively recreate and market themselves as post-industrial cities that can provide tourists and suburban visitors who have been increasingly interested in metropolitan lifestyle tourism with mosaics of playful and vibrant street cultures, unique cultural traditions, exotic ethnic enclaves, arts districts and amenities (e.g. galleries and concert halls), sports/leisure/shopping infrastructure, and (nightlife) entertainments characterized by youthful subcultures (Kearns and Philo 1993; Ward 1998). This municipal move became especially prominent from within the context of the "second wave" of gentrification in the early 1980s. Needless to say, the economy of cultural experience has been registered as a source of tax revenues for municipalities.

In addition, cultural aspects of cities were recognized as magnets to pull in culture/knowledge based post-industrial businesses, such as the art, culture, advertising, and media industries, whose presence in turn reinforces the cultural profile of cities' postindustrial businesses. Labor forces in these businesses—a sector of the new middle class, the "creative class"—have increasingly counted the possibility of a lifestyle and a quality of life, enabled by the presence of high culture, aesthetically appealing landscapes, culturally interesting commerce and amenities and authentic subcultural spaces in cities as an important factor in their choice of habitat and workplace (Ley 1996; Sorkin 1992; Zukin 1995). This is related to the cosmopolitan sensibility and subcultural savvy that these educated young professionals cultivate, sometimes as neo-bohemia (Lloyd 2006), and have developed in relation to their consumption patterns, which in turn has become an essential component of their "cultural capital" (Grazian 2005). This is the context in which Richard Florida (2004: 69–70) has been calling upon municipal governments to improve their cultural environments, and introduce a wide mix of lifestyle options and an openness to diversity, because all of this appeals to the "creative class." He even invented a "creative index" based on these criteria to evaluate the competitiveness of different cities.

The aesthetics and styles associated with neighborhoods constitute an important foundation for the re-kindling of depressed property markets too. Realtors and developers often use such aesthetics and style as a way to brand neighborhoods as chic. This enables them to reap "monopoly rent" (Harvey 2003a), in which the monopoly of the cultural distinctiveness embedded in a place enables stakeholders to reap extra rents in addition to an average differential rent (Tretter 2009: 114). It has been observed that in some cities municipalities strategically use artist communities, granting the latter with subsidies to anchor their workshop spaces in derelict buildings, with the expectation that this will revalorize the depressed property values of the neighborhood (Ponzini and Rossi 2010). Often, culture-led developments, or gentrification hinging on neighborhoods' cultural images, are also the conduits through which the investment of mega global entertainment and tourism capital is channeled into neighborhoods (Gotham 2005). Indeed, gentrification has become increasingly affiliated with the cultural qualities of cities and neighborhoods, a process I call the "culturalization of gentrification" in the book.

With respect to this process of "culturalization of gentrification," some urban theorists have celebrated the "agency" that gentrifiers have exercised

(Ley 1996). They have argued that this agency has contributed to finally restoring and liberating urban diversity and vibrancy in the central city from the 1960s formalistic modern urban planning that suppressed them and contributed to a boredom that materialized in the suburban lifestyle (Caulfield 1989). In this vein, it has been argued that gentrification is a liberating process for women and gays who can benefit from the emancipatory potential of the central city (Ley 1996; Rose 1984). Caulfield (1994) has further described gentrification as a "critical social practice" as it opens up the interaction between gentrifiers and existing residents, and increases a tolerance towards diversity. Because these urbanists saw these gentrifiers as leading the way for developers to flow into areas for gentrification, the theories advanced by this coterie of urbanists have come to be called "demand- or consumption-side" explanations of gentrification. Lees et al. (2007: 208) argues, however, that the socially positive effects of gentrification, if any, are only applicable to small parts of pioneer gentrification occurring in the 1950s and 1960s. The pioneer gentrifiers were left-liberal, new, middle-class people who were different from the conservative "old" middle classes. Inspired by the types of Jane Jacobs, these pioneers moved into the central city in a conscious pursuit of social mixing, socio-cultural diversity and mixed-use vitality.

However, later waves of gentrification actually involved gentrifiers that were different from earlier reformist gentrifiers, and spurred extensive displacement of lower-income populations and inaugurated a tide of middle-upper-class gated communities in the central city. Restored urban diversity and everyday vibrancy are now valued mainly in the interests of the middle-upper-classes and the real-estate/entertainment capitals profiting from it. Even for women, it was mostly well-educated, professional, middle-class women who have benefited from gentrification, and who are hardly representative of all women (Lees et al. 2007: 213). The unfolding aesthetic urban landscape is only a carnival mask that effectively papers over the deepening economic and social marginalization experienced by the lower strata in cities (Harvey 1988). Therefore, perhaps unintentionally, demand-side explanations of gentrification, have in effect uncritically embraced the uneven development caused by gentrification (ibid.: 211). Furthermore, some urbanists have criticized demand-side explanations, arguing that gentrification is led by capital and capitalists, developers, financial institutions, landlords and real-estate agents, rather than by the consumption desire of the new middle class; these have been called supply- or production-side explanations (Smith 1979, 1982, 1996). For example, Neil Smith has explained gentrification with the theory of rent gap which refers to the process in which, thanks to a highly depreciated property market in the central city, especially in derelict areas, the possibility of profit-making through the redevelopment of these areas is higher than other areas, and therefore, developers invest in these areas, thus triggering gentrification. Now, a consensus has been reached that supply-centered and demand-centered explanations should be combined since forces of production and consumption in the process of gentrification are always symbiotic.

In the next section, I will show how gentrifying cities that are supposedly playful and convivial have also taken on the shape of fortresses in which a militarization against "undesirable" populations is relentlessly executed. But before that, it deserves mentioning here another process by which these aspiring "creative cities" that are expected to spur cultural revitalization have instead ended up suppressing and pricing out existing creative formations, such as vernacular cultures and artist communities, and replaced them with sanitized and upscale commerce and cultural institutions (Hae 2011b; Solnit and Schwartzenberg 2000; Zukin 1989, 2009). Mele (2000) names this process "symbolic inclusion" and "material exclusion" of vernacular cultures, in the sense that the *images* of the lived, vernacular cultures and even counter-cultures of neighborhoods are appropriated and absorbed into new landscapes of middle-class lifestyle consumption, while the artist or ethnic communities that actually produced these cultures are increasingly displaced from these neighborhoods. This irony has been frequently observed in inner-city neighborhoods, or in derelict manufacturing sites, where artists have crept in as they looked for cheap and spacious living/work spaces, and developed alternative expressive cultures and lifestyle. These areas often frequently become vulnerable targets of gentrification, due, in part, to the logic that the rent gap theory explains, but also because the artistic vibe that artists create may be appealing to middle-upper-class potential gentries. Artists and subcultural communities have become victims of their own success in these cases, but this process has involved a more ironic fact. Artists often benefit from municipal and real-estate sectors' drives of cultural revitalization of neighborhoods, and they are often not concerned with the fact that their own entry into these neighborhoods sometimes contributes to the displacement of its low-income residents (Ponzini and Rossi 2010). Artist communities do not always stand together with displaced low-income victims in the latter's anti-gentrification protests, even though they often become gentrification's next victims. My book demonstrates a similar irony in the history of dance parties, dance clubs and other nightlife businesses that were located in the derelict neighborhoods of NYC in the 1970s and 1980s.

MILITARIZATION OF URBAN SPACE

The supposedly emancipatory and playful "post-modern," consumption-rich city, that was expected to propagate rich interactions between different social groups, however, has instead seen the societalization, and corresponding urbanization, of fear, as well as the emergence of new technologies of biopolitical control implemented in order to securitize urban space. The recent militarization of urban space was closely related to anti-terrorist imperatives, especially after 9/11 (Marcuse 2004; Sorkin and Zukin 2002). However, the militant urbanization which ensued after 9/11 was not an aberration specific to this particular geopolitical event, but was a continuation of the practices of neoliberal politics, under which the abrogation of conventional democratic principles, the militarization of urban space and a chill on dissent have become common to the point of being banal (Brown 2003). Conceptions of citizenship, justice and morality have increasingly been informed by neoliberal rationality, with the state justifying itself and its actions primarily according to market logic, rather than principles of democracy and justice. Issues of poverty have been ignored as a social policy objective in favor of the object of penalizing the poor, who are blamed as being "irresponsible and lazy." The rising hegemony of the "law and order" paradigm has dramatized racialized ghettos as objects of fear requiring military policing (Wacquant 2001). Migrant workers (both legal and illegal) have been another population picked out for severe policing and exploitation, having been particularly subject to sexist and racist policing practices, as well as labor exploitation by employers due to their restricted status (Pratt 2005; Sharma 2010).

The militarization distinctive to neoliberal and postindustrial societies has unfurled at urban, community and architectural scales, yielding similar results in all cases in terms of widening class disparities. First, new laws and political campaigns relating to and defined by notions of "Quality of Life" and "Zero Tolerance" have surfaced in many cities. Quality of Life and Zero Tolerance have served as instruments through which "undesirable" populations and land-uses, and the (sub)cultures associated with them, have been criminalized and subsequently displaced from gentrified neighborhoods (Hubbard 2004; MacLeod 2002; Sorkin 1992). Quality of Life and Zero Tolerance initiatives have been informed by the "Broken Windows" thesis (Wilson and Kelling 1982), in which "undesirable" populations, such as homeless people, panhandlers, addicts, yobs, sex workers, porno shops, squeegee kids, loiterers, the mentally disturbed, graffiti artists, vandals, unlicensed street vendors, drunks, gangs of teenagers playing loud music (groups often associated with inner-city black/Latino music), people of color and immigrants are, when they not outright criminalized, alleged to be symbols of "disorder" and "urban ills." They thus are understood to be triggers of urban decline as they cause negative secondary effects in neighborhoods and defile the image of the city.[4] By broadening the demographic purview of crime policing in this way, the theory of "Broken Windows" has bestowed broad new discretionary powers to policing authorities.

A plethora of literature has demonstrated how urban society and space have been disciplined through legislation, political campaigning, surveillance technologies and landscape design schemes, motifs of which resonate with the Broken Windows thesis (MacLeod 2002). The transmutation of public space into a site of experimentation for new policing technologies has been well documented and debated. Christopherson (1994: 409–12) argues that the pre-eminent concern of contemporary urban design is to manipulate human interactions, rather than encouraging the spontaneity of them, so that "normal" users of public space are separated and insulated from "undesirable" demographics. CCTV has become increasingly popular (Fyfe and Bannister 1996), causing controversies related to violations of civil liberties due to concerns over targeted policing of particular racial and sexual groups by enforcement agencies that use CCTV, and the potential abuse of information garnered. Street furniture is designed, often with a subtle code, to discourage undesired use, for example, street benches are designed tilted and with armrests that discourage sleeping on them (Davis 1992). More and more public toilets are bulldozed out of public spaces. In addition to designing out uses by the socially disenfranchised and stigmatized, new sets of laws and police campaigns have emerged as an "institutional fix" (Peck and Tickell 1994) meant to effectively regulate undesirable populations, behaviors and land uses. Sweeping "area bans" have been imposed on public spaces, in order to control criminals and drug abusers, but also to provide convenient discretionary power to police officers to oust persons of any category of "undesirability" from specific areas (Belina 2004). "Legal innovations," such as the prohibition of the mundane act of sharing food in public space, have been devised with the aim of suppressing homeless aid activists (Mitchell and Heynen 2009).

In fact, a gamut of anti-homeless legislation emerged in the Global North cities beginning in the 1990s. This range of legislation has not only outlawed aggressive panhandling, but also activities in which homeless people must engage in public in order to survive, such as sitting on sidewalks, urinating and defecating in public, loitering and sleeping (Mitchell 2003: 163). The legitimacy of these types of policing has been vigorously contested by urban poor advocacy groups. According to them, those vulnerable to these kinds of policing, such as homeless people, panhandlers and squatters, are victims of neoliberal and post-industrial changes—changes that have brought about the casualization of labor, the retreat of welfare regimes, reductions in affordable housing and deinstitutionalization—rather than being the causes of urban decline, contrary to what the Broken Windows theory argues (Mitchell 2003; Smith 1996). What the Quality of Life policing against homeless people services to do is not only to mask the structural causes of homelessness, but also to spatially displace from newly refurbished urban landscapes "undesirable," dispossessed populations. By doing so, neighbourhoods get an upgrade, bearing business friendly images and providing a sense of security for the corporate workers, middle-upper-class professionals and property owners that cities seek to attract. The security of urban space actually affects the financial security of the city, as an assessment of the level of risk of cities affects the cities' bond rates and determines the willingness of investment entry by global private capital (Mitchell and Beckett 2008: 91–95). The difficulty for the agents of urban neoliberal capitalism is that, according to Mitchell (2003: 174), homeless people are, if not wanted, still needed as part of "the reserve army of labor" whose presence is necessary for capitalists to be able to discipline the broader working class. The management of disposable people like the homeless thus comes to involve a key spatial component which involves eliminating homeless people from refurbished urban spaces, but not eliminating homelessness itself (ibid.: 167). Punitive urbanism works as a "mode of regulation" that helps to sustain the regime of accumulation of neoliberalism and post-industrialism.

In addition, changes in the normative ideals of public space, and according changes to ideals of democracy, also work as a mode of regulation that smoothes out the operation of neoliberal and postindustrial regimes of accumulation. The increasing surveillance in public space and the popular support for it has rendered public space dysfunctional, against its supposed objective as being an instance of a democratic public sphere of universal access. The freedom of movement without fear of uncomfortable encounters by a privileged class now "legitimately" trumps over the very same freedom of movement that ought, ideally, to be exercised by the under-privileged as well (ibid.: 189). This, by implication, signifies a gradual transformation of popular perceptions of rights, democracy and justice, as the nature of public space in society is often an index of these latter three. It also signifies the devaluation of an urban ideal, of the value of dynamic social interaction between diverse social groups, and hinders a learning about society by each individual that the proximity with different others would otherwise foster (Kohn 2004: 81). It would be an exaggeration to argue for a historical break occurring among cities, separating out eras of before and after the 1980s, the latter decade as marking the beginning of concerns of private property, inter-class segregation and regulation of the under-privileged "others." All of these have long been characteristic of Western cities. But, as Christopherson (1994: 410, 412) contends, contemporary cities take on more of a fortress character in the development and administration of their spaces.

The devaluation of public space and of the urban ideal is not caused only by the state, but also by the market. Growing initiatives in favor of the privatization of urban space have imposed detrimental political costs. As I discuss in Chapter 8 when examining Mitchell's (2005) case study, the privatization of once public streets has often worked to deny the rights of marginalized people to be in those streets. The public marketplace, that can in principle allow spontaneous interactions across the boundaries of different social groups and individuals, and the participation of these groups in the production of an urban public sphere, has been increasingly replaced by private shopping malls, including underground tunnels and skyways (Boddy 1992; Judd 1995), or the semi-private outdoor marketplaces of BIDs (Christopherson 1994; Kohn 2004: 81–88; Ward 2007), all of which have been built with the welcome of local governments.[5] These (semi-) private spaces have now grown as a simulacra where the nostalgic look and feel of traditional public marketplaces are mimicked, whereas ordinary activities anticipated in these traditional marketplaces, such as groups hanging out, loitering (mostly on the part of homeless people and pan-handlers), protests, community activities that deviate from, and are threatening to, consumption, are strategically displaced. As private property owners (or the association of businesses in the case of BIDs) are granted the "right to exclude," they can legitimately weed out people conducting these activities. This exclusion and displacement is executed through a code of conduct that subtly encodes who do not belong in the premises, with floor designs and the kinds of public amenities that micro-manage the traffic routes of shoppers, and private security guards (Clough and Vanderbeck 2006; Goss 1993; Staeheli and Mitchell 2006). Commentators have pointed out how BIDs symbolize the privatization of public service, as local governments, short of financial resources, tend to hand over parts of their authority, as well as their responsibilities, to business interests, ushering in serious flaws in terms of democratic accountability and equal distribution of essential services. And, this transformation of urban consumption geography may herald the normalization of "consumer citizenship" (Christopherson 1994), legitimized against a now widely suspected public citizenship.

Gated communities are another mechanism that have come to symbolize the militarization of urban space. One example is Common Interest Developments (CIDs), self-governing residential bodies in which each individual owns private dwellings, but within which facilities are commonly owned. Amenities (such as open spaces, recreational facilities, swimming pools) and roads are funded and used only by residents. These developments take on a "fortress" character (Davis 1992), with restricted entries, ramparts, and security systems. This was in response to the anxiety of the affluent over the protection of their property value from urban ills such as crime, racial/ethnic "others" and undesirable land-uses. Design movements, such as the Crime Prevention Through Environmental Design (CPTED)—of which Oscar Newman's (1972) concept of "defensible space" is paradigmatic—encourages residents to police their neighborhoods through physical design, such as gardens, window locations, plantings, landscaping, paving and signage, all of which can function as a symbolic barrier to, and as tools of surveillance over, suspicious outsiders (Christopherson 1994: 421).

Smith (1996) examines discourses that circulate in the midst of gentrification and militarization of urban space in his study of NYC. According to him, the social meaning of gentrification has been increasingly constructed with a "frontier" myth (ibid.: 13), and gentrification has often been scripted in the media as a struggle to conquer, tame and civilize the urban frontier. The frontier myth has been permuted into "revanchism" (ibid.: 211) ensuing from the stock market crash in the early 1990s and the subsequent depression and de-gentrification of the city. This has amounted to a revengeful and reactionary offensive by the middle-upper class to disenfranchised populations that some could easily identify as having been informed by a "Broken Windows" imperative. The disenfranchised were blamed for the economic crisis, as they incurred financial costs on the city by being dependent upon social assistance, sabotaging the free movement of capital. In addition to being constructed as being engaged in the "theft of the city" (ibid.: 211), they were also alleged, in popular discourses, to stand against family values, civic responsibility and neighborhood security. Smith argues that revanchism was behind the drive of punitive policing of urban space and attacks on social citizenship, both of which reached a climax under mayor Giuliani in NYC (Smith 1998), which I detail in Chapter 6. Some authors have argued that revanchism is not necessarily at the bottom of public initiatives to clean up public space; they argue that it consists simply of the efforts of authorities to make public space safer for the general public, and that even the poor have voiced their endorsement of the state's "law and order" approach (e.g. Atkinson 2003); however, this equally can confirm that the revanchist ethos has crossed class boundaries.

Zoning systems have also been put in practice with the effect of sequestering those that do not go well with refurbished urban spaces. Zoning—or spatial grids in general—has long been instrumental in reproducing social control through boundary policing according to class, racial, sexual and gender lines, as well as facilitating capitalist land markets to be exchanged in parcels (Lefebvre 1991). There is thus an inculcation of the disciplinarity of property relations in zoning system, privileging the interests of the propertied classes and of development capital (Blomley 2003; Fischler 2001). Despite the particularity of interests implicated in zoning processes, it has been deemed that zoning is set in place to promote public well-being. At the same time, thanks to the language of scientific (and often moral) rationality associated with zoning layouts, the latter have been bestowed with an "aura of legitimacy." Violations of these spatial rules, therefore, have "legitimately" been subject to state violence (Blomley 2003). For example, as Chapter 2 shows, the NYC cabaret zoning laws in the pre-1985 era that isolated the locations of black and Latino live music venues had their roots in long-standing militancy against the expressive cultures of racial others, and the fear of inter-racial mingling. The laws were justified, however, based on the science of noise (Chevigny 1991).

The moral and ideological values implicated in zoning systems, or any spatial grids, are lived in people's everyday lives, and are internalized as natural, as a norm that simply defines the relation between owners and spaces, rather than as intentionally produced through struggles between different social groups. The pre- and post-war American suburbs were engineered to be predominantly white working class, nuclear family habitats. This was enforced through combined mechanisms of red-lining practices by financial institutions that effectively restricted housing loans from black populations, and the zoning codes designated in newly developed suburban areas—such as large minimum lots or dwelling sizes—that effectively impeded the location of housing of lower property value (e.g. apartment housing), and thereupon, of disenfranchised demographics. The American suburbs are often scripted as the symbolic landscape of the "American dream," and when particular social groups are not present in this landscape, it means that these groups are also excluded from the "imagined community" of the nation. The configuration of space, zoning included, is, therefore, a code for, and enabler of, the power structure of a society. Geometry, spatial grids and geographical boundaries are programmatic products of power relations, however accidental they may look.

(Re)zoning has also been the site of popular struggles as cities have used zoning systems in de-territorializing and re-terrotorializing urban space to accommodate the newly preferred best land-use for maximum profitability, in accordance with the imperatives of changing political economy and preferred demographics (Feagin 1998). The same is true with regards to zoning processes in the era of neoliberalism, post-industrialism and revanchism. Municipalities have expedited gentrification by granting zoning variances to real-estate developers and property owners, and by eventually up-zoning devalorized areas. As Chapter 3 shows, the rezoning of SoHo that allowed residential developments—which were ironically the fruits of activism of squatting artists—paved the way for massive loft-living developments, at the end of which artists were ironically displaced from the neighborhood. Some scholars have proposed public space zoning that controls chronic "misconducts" in public space, such as panhandling and bench squatting by homeless people, in order to protect respectable people (e.g. Ellickson 1996; for a trenchant critique of this argument, see Mitchell 2003: 211–18). In the revanchist NYC of the mid 1990s, mayor Giuliani revised zoning regulations in order to zone out sex-related businesses, such as porno shops, from gentrifying neighborhoods, as is discussed in Chapter 6.

On the other hand, Ranasinghe and Valverde (2006) argue that municipalities are fundamentally limited in their means of implementing social policies, due to the downloading of social service responsibilities from higher levels governments, without accompanying funding resources. They therefore tend to resort to zoning mechanism to solve social problems—for example, by amending zoning codes to provide homeless shelters in certain areas. However, these types of municipal attempts often have to confront the NIMBY politics of property owners. Despite the celebratory overtones over the participatory urban planning paradigm that has gained a great deal of popularity since the 1960s, land-use planning has actually resisted democratization over the same period (ibid.: 328). Rights in land-use are primarily tied to property, and groups that influence the land-use planning are mostly the propertied (and to some degree, those who have legal occupancy, such as tenants) that would be provoked at any possibility—such as homeless shelters in their neighborhoods—that degrades property value and their quality of life. This is why zoning provisions are often opposed to the kinds of land-uses that are related to social justice. As I show in detail in several following chapters, the cabaret law zoning regulations have also been the subject of this type of NIMBY politics.

THE GENTRIFICATION OF NIGHTLIFE

Nightlife businesses, especially the ones associated with inner-city subcultures, often constitute an important foundation for the re-kindling of depressed property markets in derelict neighborhoods, by bringing people to abandoned streets, adding a particular aesthetic to the image of the neighborhoods, and by helping to generate a vibe and image of lively urban socialization. Realtors and developers often use such aesthetics and style as a way to brand neighborhoods as chic. Florida (2004) also argues that nightlife is an important component of the mix of lifestyle options appealing to the creative classes, most of whom are young workers. "A vibrant, varied nightlife," Florida's findings demonstrate, "was viewed by many as another signal that a city 'gets it,'" and some of his research subjects even complained about certain cities where the nightlife ends too early (ibid.: 225). Though, Florida's research subjects displayed, among nightlife options, a lower preference for "bars, large dance clubs and after-hours clubs" than venues closer to "cultural attractions" such as venues for symphony, theater, (jazz) music, late-night dining and coffee shops (ibid.). On the other hand, Currid (2007: 87–113), another champion of creative classes and economy, underlines the importance of nightlife businesses like bars, lounges and large dance clubs (in her case, in NYC), as these sites provide spaces where young urban creatives in fashion, advertising and music industries casually hang out, exchange information, and form an informal social network which, she argues, contributes to the creative cultural economy of the city. For these reasons, nightlife has received renewed attention by policy makers as a driver of post-industrialization and urban revitalization. Nightlife businesses are now less associated with an "immoral" underworld or dangerously liminal and transgressive activities, than they used to be, and are instead looked upon as a legitimate industry that supplies post-industrial "lifestyle consumption" goods to cities and enhances cities' images as lively, cosmopolitan urban habitats (Hobbs et al. 2005).

While Florida's proposal has attracted the attention of many North American policy makers, the recognition of the significance of nightlife as an appealing feature of contemporary cities has taken on the corporeal forms of legislation and political initiatives in Britain (for an overview, see Shaw 2010). In sync with the deregulatory trends of neoliberalism, the UK government legislated a series of restructured (de)regulatory systems for nightlife since the 1980s, the most prominent example of which was the 2003 Licensing Act legislated under the Labor Party of Tony Blair. The deregulatory measures included an extension of licensing hours, release of the restrictions on the number of permitted nightlife establishments, and an introduction of a more systematic and standardized decision-making process in granting licenses to nightlife businesses in order to discontinue arbitrary and moralistic decision-making processes (Talbot 2006: 160). Such measures were expected to encourage nightlife, particularly in impoverished cities, such as

Manchester, Leeds and Newcastle, in order to develop it as an economic resource (ibid.), as a landscape to provide a vibrant image of the cities (Ward 1998), and as an important catalyst for neighborhood gentrification projects (Talbot 2006: 163).[6] Such campaigns were variously labeled under headings like "Night Time Economy (NTE)" (Bianchini 1995; Lovatt and O'Connor 1995) and "24 Hour City" (Heath 1997).

In North America, there has been no equivalent in terms of comprehensive national initiatives for the promotion of nightlife, as has been the case in Britain. Nonetheless, North American cities have also promoted nightlife to attract tourists as well as new residents to downtowns abandoned due to suburbanization since the 1950s and the fiscal crisis of the mid-1970s. Initiatives like "entertainment districts" or financial and legal support for flagship projects like large-scale corporate entertainment destinations have become popular (Campo and Ryan 2008; Gotham 2005). Even in the case of neighborhood nightscapes that, unlike UK counterparts, have been formed without direct governmental sponsorship and are generally non-corporate, liquor agencies at the state governmental level have often encouraged the proliferation of nightlife businesses (bars, clubs and lounges) as part of urban growth initiatives through permissive license-issuing practices (Ocejo 2009: 9).

While policy makers may give a nod to the promising role of nightlife in enhancing the city's image and economy, there remain quite a number of concerned voices that point out how the deregulatory policies would increase the harms and risks associated with nightlife in urban society, such as "binge drinking," alcohol-related violence, vandalism and disorder, noise nuisances, immoral and improper subcultures (Hadfield 2006; Hobbs et al. 2005). In North America, conflicts have been prominent in cities like Philadelphia, Cleveland, Tampa, Seattle, Toronto and Austin, where municipalities and real-estate developers have encouraged condo developments and loft conversion near entertainment zones, ironically marketing to potential buyers the vibrant ambiance of the areas created by the presence of nightlife (Bhatt 2007; Campo and Ryan 2008). Nightlife tends to develop and proliferate with gentrification (Ocejo 2009: 8, 10) and is often encouraged to locate in gentrifying neighborhoods by developers, landlords or state/municipal governments, so gentrifying neighborhoods have soon been inundated with various nightlife businesses to the dismay of some residents newly moving into these neighborhoods. The "nightlife fix" to urban economies, spaces and images in decline, backfired, therefore, with a resultant dilemma that pressured municipal governments to carry out contradictory policies of re-regulating nightlife businesses while de-regulating them at the same time. Various laws have also been employed in North American cities to administer the nuisance effects of nightlife (Berkley and Thayer 2000; Bhatt 2007), often in conjunction with broader Zero Tolerance policing and Quality of Life initiatives prevalent in neoliberalizing cities. Institutions like the "Responsible Hospitality Institute" consulting firm (http://rhiweb.org/) emerged to provide government officials and nightlife entrepreneurs with standardized toolkits about how to create safe and nuisance-free, but nightlife-rich, sociable cities, in collaboration with residential communities.

In Britain, the Labor government also responded to concerned voices over the negative impacts of the Licensing Act, stressing responsibility on the part of drinkers and business operators. However, despite this response, scholars studying nightlife issues (e.g. Hobbs et al. 2005: 169–70) argue that nightlife governance (both over nightlife premises themselves and neighboring public spaces) was characterized essentially by the withdrawal of the state, consistent with neoliberal ideology. The state has rendered the growth of night-time consumption to proceed unchecked, allowing pubcos to sell large quantities of alcohol at low prices through chaotic implementation and poor enforcement of licensing policies, which in turn has encouraged widespread excessive drinking, and has built into criminogenic nightlife conditions (Hobbs et al. 2005; Roberts 2006). What emerged to ensure the "orderliness" of nightlife then was an increasing reliance on private security, such as bouncers and door staff, who tend to resort to intimidation and violence. The central government strategy in terms of nightlife may be about to change substantially, as the current Conservative/Liberal Democrat coalition are seeking to nullify the Licensing Act to tackle these problems (Johnson 2010).

On the other hand, Talbot (2006: 160) argues that as concerned voices over nightlife problems rose, "new forms of social and cultural differentiation" have come into being at the local level, as a way to distinguish what is acceptable/unacceptable, and what counts as orderly/disorderly premises. It has also been observed that such perceived differentiations between orderly/disorderly venues quite often overlapped with the class (Chatterton and Hollands 2003), or racial and ethnic composition of these venues (Böse 2003; Talbot 2004). In Talbot's case study (2006), she observed that among key authority figures in the licensing regime, the perceived distinction between orderly and disorderly venues was also often layered with the notion of "commercial viability or business competence", aligning "the ability to maintain order" with the economic capacity of the venue (ibid.: 164). The venues catering to Afro-Caribbean patrons were generally considered to be lacking in business capacity, and in such an equation, they were easily associated with "disorder," and subjected to more tightened surveillance and penalization. In contrast, it was mostly profitable and commercial venues, such as large chains, or a syndicate company that established dance bars catering to young white folks that were considered to be efficiently run and capable of maintaining order.

Hadfield (2008: 441–42) shows that governmental officials in charge of licensing, and the judges in the licensing courts, tend to buy into the arguments presented by the upmarket corporate night businesses that their economic capacity guarantees them the ability to maintain proper security.

And, when these upmarket businesses try to prove their security credentials, they tend to stress the particular racial/class make-up of the patrons of their businesses (using a variety of practices, these clubs tend to exclude racial minorities from their venues), thus tapping into the preconceived notions of the licensing authorities about what racial/class venues count as orderly and disorderly venues (Talbot 2006: 165). Under such logic, profitable and commercial venues received most of the financial and legal assistance from the municipal government available through gentrification initiatives (ibid.: 166). This, combined with the general fear of Afro-Caribbean venues among young, white, professional gentries (and nightlife patrons), and high rents resulting from neighborhood gentrification that these venues cannot afford, lead to the marginalization of these Afro-Caribbean venues in the neighborhood (ibid.: 165–66). Smaller local/alternative/independent nightlife venues cannot compete with corporate/upmarket sectors that are extremely well industrialized with substantial capital investment, rationalized production techniques and well-established connections to specialist law firms. The smaller, non-corporate venues are thus priced out of central city locations (Chatterton and Hollands 2003: 19–44; Hadfield 2006: 47–48).

Members of these alternative nightlife sites sometimes become mobilized to protest against governmental crackdowns on these sites and to further campaign to make claims for people's rights to party, and against rote corporate infiltration into the fun, leisure and creativity of urban life. For example, as governments moved to make laws that meticulously targeted rave parties, these rave formations have been rapidly politicized and become locations of public protests (Huq 1999). However, Chatterton and Hollands (2003: 60) state that the organizing drive among local/independent/alternative businesses has been weak or short-lived and, consequently, that they have been under-represented in political and judicial fields. On the other hand, corporate and gentrified nightlife have formed powerful trade organizations (operated with their overwhelming financial resources) in order to lobby politicians to act in favor of their interests. The gentrification of nightlife, thus, has been re-inscribed by unevenly distributed political capital between different sectors of nightlife.

Talbot (2006) argues that all these processes have led to a "subcultural closure" (159), as forms of "wilder," experimental and culturally diverse (and arguably more important), nightlife get increasingly purged out of gentrifying neighborhoods, and are replaced by gentrified forms of nightlife (165; see also Hadfield 2006; Chatterton and Hollands 2002). In the gentrified streetscapes that seemingly represent the ideal of 'authentic' mixed-use neighborhoods, the wilder version of nightlife often only remains as an image, as a simulacrum of the neighborhood's sub-cultural history, wherein communities that produced this wilder nightlife and its spaces cease to exist. This again shows how Florida's 'creative city' type of policy development can ironically turn destructive towards creative sub-cultural formations in cities. This also has the important consequence in social and cultural life in contemporary postindustrial cities, as citizens' access to—and by extension, their right to—diverse and experimental urban subcultures becomes increasingly limited. The analysis of such transformation of nightlife under gentrification is, therefore, an important extension of other relevant research that have brought to light the marginalization of certain populations under gentrification and Quality of Life policing, such as homeless people, together with the implications of such marginalization for the declining principle of the "right to the city," in the form of citizens' rights to social housing and public space. Patterns of the transformation of nightlife have varied across different locales, and have been contingent upon the specific political economic directives, institutional and legislative landscapes, licensing, planning and enforcement systems and cultural and social make-ups within which each has been historically formed (Jayne et al. 2008; Jayne et al. 2010).[7] Despite divergences, however, there have been observable trends of "gentrification of nightlife" and "subcultural closure" across different locales.

As I present in this book, a process similar to "subcultural closure" and the "gentrification of nightlife" was also observed in NYC, as the city has been extensively gentrified and, simultaneously, new political initiatives promoted by the anti-nightlife groups have been mobilized to strictly police and even displace (certain types of) nightlife. In addition, here, a process of what I call "gentrification with and against nightlife" (Hae 2011b) in several neighborhoods was also prominent. I will also show in Chapter 7 how the gradual gentrification of nightlife and uneven development within nightlife in the city have shaped, and been shaped by, particular patterns of pro-nightlife politicization, and the power struggles between pro-nightlife actors that are situated differently in relation to the gentrification and post-industrialization of NYC. This also shows why the politicization has been limited in defending democratic nightlife from corporatization and governmental over-regulations. Based on this analysis, I argue in this book that looking into the politicization of nightlife and the struggles that take place between different fragments of nightlife can enrich and complicate our understanding of gentrification and post-industrialization. Studying the politicization of nightlife also provides a venue in which to muse over the transformative potential of nightlife. Nightlife's transgressive and dissident potential has long been recognized by historians and cultural theorists (e.g. Palmer 2000), but this recognition should be revised and contextualized in relation to new urban milieus of neoliberalization, gentrification and the state's re-regulations of nightlife. Despite this need, and also despite the nascent politicization of nightlife in various cities (for example, in the form of organizations such as the San Francisco Late Night Coalition and the Seattle Nightlife and Music Association), few have actually studied the complicated, fragmented and contradictory processes of nightlife politicization and power dynamics within these formations, a key exception being Chatterton and Hollands' (2003) insightful, but still schematic, analysis. My empirical case in this book seeks to fill this missing detail of nightlife studies.

THE RIGHT TO THE CITY

The story of this book calls for attentions to how urban milieus shaped in the midst of gentrification and the emergence of Quality of Life policing have also undermined social conditions necessary for people to be able to participate in the production of spaces of use value. In this book, I make a claim that it is critical to restore these conditions, and that the notion of the "right to the city" (Lefebvre 1996) bestows a political and legal as well as theoretical platform useful in developing popular struggles for the reclamation of these conditions.

In establishing the grounds for this discussion, I want to first examine Lefebvre's (1996) conceptualization of "the urban." "The urban" for Lefebvre does not so much have to do with the boundaries of a city *per se* as having to do with the *process* of social/spatial (re)production of (counter-hegemonic) power (Dikeç and Gilbert 2002: 65). "The urban" constitutes resistance against "the irreversible tendency towards money and commerce, towards exchange and *products*" (Lefebvre 1996: 66), against "managed consumption" (ibid.: 147), against the life that bureaucrats, planners and capitalists design, develop and provide for people. Lefebvre claims that the urban is an *oeuvre*, a work of art, and as such, involves both use value and *la Fête* (ibid.: 66 and *passim*). The urban is not something organized by the state or market, but is realized through citizens' appropriation of, and participation within, the space and time of the city for the purpose of "creative activity" (ibid.: 147) and "play" (ibid.: 171), those not primarily defined by state or market ideologies. "The urban" is as much about the process of radical participatory democracy in urban production and management, and the accordant end result of a society alternative to what Guy Debord (1994), another architect (with Lefebvre) of the Situationist International in France in the 1960s, called "spectacle," which is a society of commodification, alienation, homogenization and boredom.

Lefebvre further (1996: 75, 109) stressed "the urban" as ensembles of differences, social life featuring encounters with difference in everyday life (Also see Duke 2009; Lefebvre 1991). The attention to differences is also found in many other scholars' work. For example, according to Young (1990: 240), city life symbolizes "a being together of strangers, diverse and overlapping neighbors," which she defines as one of the primary sources of social progress related to cities. She maintains, therefore, that strategies that seek to bring social justice to cities should take seriously "the realization of a politics of difference" as "the normative ideal of city life" (ibid.). As one condition in accomplishing this politics, urban space should be constructed to support the diversity of activities that exist in cities (ibid.: 239), and therefore, function as part of a system that guarantees democratic access to urban space for diverse groups and individuals, and opportunities for them to appropriate and explore urban space for pleasure and excitement (ibid.: 239–40). To be sure, the assertion in favor of diversity for Young and Lefebvre is one that does not put diversity within the sphere of market competitiveness and state control, as multicultural policies of many Western cities have done. It is more about coupling radical participatory democracy with political and economic equality to every social group and individual.

The normative ideal of city life, however, has been constantly under attack, not only by the disciplinary forces of the state, but also by market forces that tend to colonize the use value of urban life. Mitchell (2005) analyzes how the privatization and militarization of urban space in contemporary cities have diluted people's "rights" and shrunk the possibilities of "the urban." Mitchell shows this with a legal case in which, as once-public streets were privatized, particular "undesirable" people, such as a young black male involved in the case Mitchell elaborates, were forbidden to be on these streets. To (be able to) be present in urban public spaces, Mitchell (ibid.: 582) maintains, is a "crucial precondition" for the "right to the city", as it provides the chance "to be visible as a member of the urban public," and is thereby a basic condition for a democratic public sphere. However, the right to be in a public space is compromised in the face of changing geographies brought about by privatization, which has led to the decline of public space, and, accordingly, that of the democratic ideal in cities, as I have detailed previously. The courts have equally been responsible for this decline. Judges in the case of Mitchell's analysis refused to grant the activity of "being on the street" with constitutional protection, as they refused to associate it with "freedom of speech," which would have granted it protection under the First Amendment. These judges also divested themselves from engaging with the problems that broader socio-economic processes, such as the privatization of urban space, have caused in urban life, that is, the process in which the status of the street has been invalidated as a public forum, and subjected people like young black males to arrest when they wind up, now, trespassing on newly privatized streets (ibid.: 579).

This approach taken by courts, which Mitchell (2005) terms "judicial anti-urbanism," betrays how liberal legalism is limited and even flawed as a forum in dealing with urban problematics that inform struggles in everyday urban life. "Rights" are still useful tools in curtailing the state's abuse of power and to redress urban problematics, but rights within liberal legalism, rooted in the historical moment of the 1789 Declaration, cannot properly deal with a gamut of new challenges that face urban society (Fernandes 2007: 207). The constitutional protection of rights is primarily limited to civil/political rights—such as (narrowly defined) rights of free speech, expression and assembly. This explains the emergence of the "right to the city" movements as an alternative both in the Global North and South, in which geographies of social justice and human dignity are sought in the form of rights to land, social housing, public transportation, public space, broader citizenship, democratic urban planning, participatory budgeting, regularization of slum settlement and so on (Fawaz 2009; Harvey

2008; Mitchell 2003; Purcell 2003)—rights that liberal legalism does not properly address, let alone champion.

However, is the normative framing of new movements around "rights"—which itself is essentially a liberal legal category—a desirable move when seeking to transcend the confines of liberal legalism? Rights have suffered skepticism among scholars that have taken social movements and social change seriously. Some schools of Marxist tradition have long expressed mistrust about rights, insisting that rights, like other liberal legal categories, effectively cater to particular bourgeoisie interests by presenting a rhetoric that misleads us into believing in the possibility of universal equality under liberalism, promote individualism, and often stand in for social emancipation while displacing the need for class struggles among social actors (Lukes 1985, 1991: 182).[8] Rights tend to reify more complicated social processes and social relations (Tushnet 1984), pushing activists to zero in on one reified issue, focusing, for example, on pressuring the state to provide sufficient "social housing," rather than fighting to correct the more structural mechanisms of capitalism—the realm of production—that engender uneven development, which manifests itself with increasing numbers of the homeless and the precariously housed (Brown 2003). Activists often fetishize, further, the single entitlement encapsulated within a particular "right" as something that will bring about significant social change (ibid.). Rights—whether they are traditional liberal rights or social/economic rights—depend "for their implementation on the availability of resources" (Lukes 1991: 176), require institutional systems of enforcement and, most of all, a will to implement on the part of the state. Without this, therefore, rights are of disutility, and can even be counter-productive. Whether rights are translated into actual social progress or not is contingent upon geometries of power and the likelihood of various circumstances taking place at a specific historical and geographical juncture.

Scheingold (1974) also argues that rights, as well as litigation and laws in general, should not be counted on as agents of change. However, he goes beyond the skepticism of rights:

> Instead of thinking of judicially asserted rights as accomplished social facts or as moral imperatives, they must be thought of, on the one hand, as authoritatively articulated goals of public policy and, on the other, as political resources of unknown value in the hands of those who want to alter the course of public policy. The direct linking of rights, remedies, and change that characterizes the *myth of rights* must, in sum, be exchanged for a more complex framework, the *politics of rights* . . . (6–7, emphasis in original)

What Scheingold means by the "politics of rights" is, *inter alia*, the capacity of rights, rights talk and rights-based litigation "in fomenting change when used as an agent of political mobilization" (ibid.: 9). It has been observed in many social movements that rights talk is instrumental in arousing (new) consciousness of particular citizen-based entitlements that have not yet been conceived of as such. As Chapter 4 in this book shows, live music musicians in NYC in the mid 1980s started to frame their work in terms of rights (i.e. live music as free expression), and this rights talk offered a critical momentum in the mobilization of musicians themselves in the sense that it helped to provide a new consciousness that valorizes their own work as an essential right, to be protected from sweeping municipal regulations. A range of the "right to the city" movements that burgeoned in the 1990s onwards have also used the discourse of rights as an organizing principle, as a "working slogan and political ideal" (Harvey 2008), to demand what has been thought "impossible," to shape a popular consciousness about the due entitlements of urban inhabitants that have been denied to them, and translate this consciousness into an action. Rights talk can gain traction more easily, and work as a lingua franca that constitutes a new collective identity and sense of solidarity among grassroots with different positionalities across different locales (Mayer 2009: 367).

Movements proclaiming the "right to the city" have been seen by urbanists like David Harvey and Peter Marcuse, and even Lefebvre himself as an oppositional demand, and less a legal claim enforceable through a judicial process (Mayer 2009: 367). However, one may wonder why we can only conceive the concept in this circumscribed sense. As Mitchell (2003: 29) argues, rights are more than a tool for popular mobilization. Once rights are entrenched into law, it provides a valuable institutional legitimacy and support to grassroots groups in their struggles, and the legal ground through which to pressure state authorities to comply with what these rights mandate. And, when these rights are infringed upon by an authority, the authority often loses its legitimacy and triggers popular backlash (see also Harvey 2000; Thompson 1975). As well, institutionalization can help to gradually normalize into popular recognition a range of entitlements hitherto unknown and unthinkable as legitimately inalienable rights.

Cases do exist where the principle of the "right to the city" has been institutionalized. For instance, Brazil legislated the City Statute in 2001 based on the "right to the city" principles, and entrenching it into the federal constitution. This is expected by some to introduce a new judicial as well as political paradigm, in unprecedented forms of social rights to land and grassroots participation in urban policy planning (Fernandes 2007). There has also been a global mobilization since 2003, joined by UNESCO and UN-HABITAT, to develop and adopt the "World Charter on the Right to the City." Mayer (2009: 367–69) argues that these institutionalizations, while having opened up an empowering space for grassroots movements, have actually raised an array of thorny questions. The Charter has mainly been proposed as a guide and toolkit for "good governance" for municipalities and NGOs. Also, the "right to the city" has settled as a smart brand with which international NGOs, for whom rights-based claims have taken on the central place of their activism, facilitate their collaborations with states and corporations, and secure sponsorship of UN organizations and the World Bank. From the perspective of NGOs and municipalities, the "right to the city" often refers to empowering civil society and through this, enhancing local competitiveness and endogenous developmental potential. This turn of what the "right to the city" signifies will effectively immure grassroots' pursuit of social justice within state-sanctioned and market-aligned public forums (Fawaz 2009: 832). It renders much less potent the radical potential of the "right to the city" that originally characterized Lefebvre's arguments (Mayer 2009: 369), which does not merely mean right of access to what already exists. It is, rather, a demand for grassroots empowerment towards the establishment of radical participatory democracy and social justice *with* transformative power (Harvey 2003b: 939)—power that effects "impossible" changes, a change, for example, from market-oriented social/spatial relations to non-market focused ones.

The capture of the "right to the city" discourse by mainstream institutions to the effect of de-politicization only increases the urgency by which grassroots need to re-claim this right, with radical ideals. Rights as a resource of bourgeois rule are not pre-determined; rights, as a socially constructed category, are subject to contestation and, potentially, a more radical re-conceptualization (Bartholomew 1990: 258; Boyd 2009: 586). Rights can be conceptually expanded from the confines of liberal legalism to include the protection of people's economic and social rights *against the market* as well as against the state, and this is what a more radical edition of the "right to the city" has worked on. The "right to the city" need not only be a negative right; it can be made a positive right with which grassroots can call for pro-active action by the state to provide socially egalitarian and accessible social/spatial policies. Rights that supposedly promote individualism can also be re-configured to promote common goods and solidarity among different individuals and social groups. The history of social movements has proved that liberal institutions and laws are not exclusive to the interests of the liberal state, the ruling class and the propertied, but have also been appropriated and re-occupied by the subaltern as sites for their struggles and for social progress. To deny this would be to "throw away a whole inheritance of struggle *about* law" (Thompson 1975: 266, emphasis in original, cited in Boyd 2009: 599).[9] Again, it is important not to mythologize rights as an agent for real social change. But, it is still more important to conceive of rights (institutionalized or not) as a pre-condition—a necessary but not sufficient condition—in the long-term and broader social movements to achieve a democratic and egalitarian urban society.

As a political project that seeks to achieve a radical participatory democracy, redistributive justice and autonomy unhinged from market logic and liberal ideologies, the "right to the city" movement faces a challenging path ahead, as it is situated within the world of vested interests of urban land markets and economies holding formidable power under the aegis of liberal legalism. This simply underlines all the more why we need a political mobilization appropriating the symbolism of the "right to the city," in which naturalized capitalist relations and the taken-for-granted social hierarchy between haves and have-nots, and between the propertied and the propertiless, and more specifically, the TINA discourse that justifies gentrification and privatization as the only viable urban policies, are questioned and challenged, and people's collective rights to use value in urban life and space are inscribed as a norm. Contrary to the traditional framework of rights under liberal legalism, the "right to the city" helps us to rethink the material conditions and social relations that we live within, and what necessary entitlements are lost in the midst of the naturalizing forces of capitalism. The "right to the city" helps us to pursue the normative ideal of contemporary urban society—the ideal that should structure and guide social life in cities.

SOCIAL DANCING, NIGHTLIFE AND THE RIGHT TO THE CITY

Now, I want to discuss the "right to the city" in relation to our rights to (spaces for) social dancing and nightlife. Mitchell (2005: 580) defines the right to be in a public space as an example of "urban rights" that enable what Lefebvre called "the urban." While Mitchell does not detail what he means by "urban rights," I interpret it as denoting a set of sub-categories that constitute the broader principles of the "right to the city"; that is, if the "right to the city" is an abstract concept that puts as its normative ideals the rights to participation, appropriation and diversity, "urban rights" can be defined as a set of specific necessary conditions that enable the broader and more abstract principle of the "right to the city." Drawing upon Lefebvre, "urban rights" can be defined as rights that open up democratic access to multiple urban spaces (including spaces of play and for use value), and democratic participation by diverse individuals and groups in the production of urban space; that enable a diverse cultural life and socialization unique to cities; and that enable (geographies of) just relations of production and social reproduction. These conditions would bring us an organic vibrancy unique to urban societies, would empower grassroots fights against discrimination, homogenization and commodification of urban life and lived space, and would, thus, lay the foundation for the progressive potential of cities.

In concrete terms, the right to socialized ownership of means of production, the right to a living wage, affordable housing, affordable public transportation, water, food, a clean environment, citizenship and the like, can compose the ideal urban conditions that enable the (geographies of) just relations of production and social reproduction mentioned above. Urban rights, as they are related to democratic access to urban space by diverse individuals and social groups and the production of geographies of diverse cultures and social interaction, would involve, for example, the right to public space. More specifically, the right to secure sufficient public space counter to the encroaching privatization of urban space, and also the right to democratic access to public space for the purposes of hanging out with peers, encountering strangers and randomly starting conversations with them without being discriminated due to one's racial/gendered/ sexual and/or other ascriptive identities (as discussed in Mitchell 2005). Rights to mundane urban activities such as associating with others to play physical and recreational activities and to cultural ceremony (especially for minority, immigrant groups) without undue restraint, and the right to secure sufficient spaces for these activities are, if seemingly trivial, banal and insignificant, nevertheless the *sine qua non* of urban rights, that make the experience of urban life authentic, diverse and democratic.[10] These are, needless to say, also the pre-condition that allows space for political activities closely associated with well-established constitutional rights such as free speech, expression and assembly.

The story of this book engages with the question about how the possibilities of realizing urban rights, registered through nightlife, and in particular through social dancing establishments, have been arrested in a gentrifying and post-industrializing New York City. Like countless mundane activities and spaces for them that play basic, but important social, cultural, and political roles in urban society, social dancing and its spaces—bars, lounges and music/dance clubs—too provide opportunities of unique ways of communication, socialization, and an experience of collective joy. As is examined in Chapters 3 and 4, social interactions mediated through social dancing and broader nightlife subcultures took place in NYC among groups very specific to urban society, such as youths from racial and sexual minority groups in NYC in the 1960s and 1970s (Fikentscher 2000). The carnival, or *la Fête*, that, according to many, has disappeared from civilized western society (Ehrenreich 2007b), was revived in the spaces in which these communities gathered and celebrated their identities through music and social dancing. Spaces for social dancing and broader nightlife constituted and promoted "the politics of difference" in Young's phrase, and in a Lefebvrian sense, these spaces also symbolized "use value" for this group of people, created through the appropriation of (derelict) parts of the city, and culminating as their *oeuvre*. These spaces of fun and pleasure have also become important sites for the evolution of realpolitik for the civil rights movement, and as in the case of some of the contemporary youth rave movements, subcultures premised upon the belief in social dancing and parties have also laid the foundation for progressive street politics with an alternative philosophy of lifestyle, environment and public space (Jordan 1998; Klein 2000: 311–24). As various chapters in this book detail, spaces for social dancing and nightlife have also warranted citizens' access to diverse expressive arts because so many expressive arts have been nurtured in these spaces. The right to social dancing is, therefore, an important urban right that enriches the cultural as well as the social and political lives of cities.

The changing geography of NYC—i.e. the economic, social and cultural geography transformed by the increasing gentrification of urban space, the post-industrialization of the urban economy and the neoliberalization and militarization of urban social life—has, however, rendered the realization of these urban rights as violations. The cabaret law was enforced in a sweeping manner to protect the property rights and the quality of life of new residents by repressing spaces of nightlife, especially spaces of social dancing. The effect of the regulation was nothing short of killing a vital urban nerve in the city, as Chapter 6 details. Repressing spaces for social dancing has become equal to denying the opportunities to people the right to enjoy a unique method of socialization, which develops the fundamental human nature of dancing into collective pleasure, ritual, and occasionally politics, all of which are important qualities of "the urban." The courts fraternized with the municipal state in applying the cabaret law. As shown in the *Festa* decision which involved the test of the constitutionality of the cabaret law (which is detailed in Chapter 8), social dancing, like many other mundane urban activities, was not recognized as a form of expression under the courts' "categorical approach" in the matter of constitutional protection (Chevigny 2004; Cole 1999), and therefore, the governmental regulation over spaces for social dancing through the enforcement of the cabaret law was (and is) only subject to a relaxed test of its legitimacy. The courts withdrew themselves from discussing the deleterious repercussions of gentrification which led the municipality to abuse the cabaret law in governing nightlife. In this way, questionable punitive policing by the state often passes the legitimacy test, and has gone unchecked by the courts. This is another example of what Mitchell characterizes as "judicial anti-urbanism," and how limited liberal legalism is when it comes to the protection of diverse urban activities from the abuse of state power.

I propose in this book that the hitherto unacknowledged value of social dancing and its worthiness of constitutional protection can be better framed, claimed and salvaged in its articulation with the normative ideals of "the urban" that the principles of the "right to the city" and "urban rights" aim to establish. Considering the contributions that rights talk makes in arousing a new political consciousness about unknown values, as discussed earlier, the claim for the protection of spaces for social dancing as a part of the "right to the city", that enables diverse and vibrant urban social/cultural life, may gain traction in transforming popular perceptions of the (hitherto unrecognized) importance of mundane urban activities, associations and spaces, as well as spaces for social dancing themselves. And, as the ensuing chapters describe, the same can be argued in favor of the protection of spaces for broader nightlife activity. Nightlife has been perceived as amounting to a noisy and boisterous nuisance threatening the quality of life of neighborhoods, so groundless governmental crackdowns on nightlife have not only been taken-for-granted as just and fair, but have even been highly acclaimed. This has undermined one vital source of "the urban," as noisy, boisterous, zany and diverse nightlife is an important context as a playground for diverse expressive cultures, unique social interactions and radical politics. The state's repression of spaces for nightlife as well as social dancing should thus be subject to strict scrutiny and a thorough probing in terms of its legitimacy.

CONCLUSION

In the course of this volume I investigate how neoliberal and post-industrial urbanization in general, and gentrification in particular, has necessarily entailed struggles over changing cultural geographies (e.g. geographies of nightlife) and simultaneously, over urban inhabitants' legitimate rights to spaces for mundane social activities. The book delivers the details of these struggles, conveying the voices of different groups pursuing different values from within urban life, cultures, spaces and rights. Besides explaining a specific urban struggle surrounding spaces for social dancing, this book also makes a normative conclusion drawing on Lefebvre's radical concept of the "right to the city". I contend that ordinary urban activities such as social dancing and other nightlife activities and spaces for them are foundations of accomplishing the normative ideals of urban life; and that the state's ungrounded repression and the market's colonization of them need to be taken seriously and strenuously opposed. Looking into the ways in which spaces for social dancing are disputed in NYC, and to what effects, enhance the understanding of how we have lost, under capitalism, the possibilities inherent to radical participatory democracy and the transformative power of the grassroots, and how we can claim them back from the disciplinary power of the state and the market—not only by taking back people's right of democratic access to social dancing and nightlife, but also more broadly, by reclaiming people's right of participation in the production and reproduction of urban life and urban space, in the direction of justice, equality and democracy. This is the intervention and contribution that this book makes in relation to the scholarly discussions of transformations of urban space and life in contemporary cities that have been wrapped up with the marketization of notions of justice, democracy and rights, and the political movements that have hinged on the notion of the "right to the city" as they have fought the nefarious effects of these transformations in response.

2 The Cabaret Law Legislation and Enforcement

In this short chapter, I introduce the institutional matrix and administrative procedures within which amendments and enforcements of the cabaret law have been carried out in NYC. I delineate what institutions and organizations have been involved in amending and enforcing the cabaret law, and to what effects. This information will help to contextualize legal and political struggles that have been unraveling since the 1970s over the cabaret law and nightlife in the city, and the unequal power relations that have taken place between different actors involved in these struggles.

The cabaret law was originally created during the Prohibition era to control speakeasies, and later developed into a tool to impose zoning and licensing regulations on the city's nightlife entertainment businesses (Chevigny 1991). The cabaret law regulation of the city's nightlife entertainment businesses until 1990 largely bifurcated along the lines of two types of entertainment uses. Eating and/or drinking places that offered "incidental musical entertainment"—music played on records or jukeboxes, or music played by not more than three musicians that did not include percussions and/or horns—were subjected to the same degree of licensing and zoning regulations as ordinary restaurants and bars. These businesses were categorized as Use Group 6A. Once the business allowed broader entertainment, that is, if the business allowed its patrons to dance, or had more than three musicians and/or allowed musicians to play horns or percussion, these venues were subjected to much more stringent forms of licensing and zoning regulations. These businesses were categorized as Use Group 12A, and were called *cabarets* in administrative nomenclature. Due to the provision that limited the number of musicians and the type of instruments, the cabaret law effectively discriminated against live jazz music played in bars, restaurants, and clubs.

Multiple regulations differentially applied to these two types of entertainment. Among these regulations, however, the zoning regulation is central and has also been the most controversial in historical struggles over the cabaret law. The zoning resolution that regulated the city's entertainment businesses until the 1990 rezoning of the cabaret law was as follows: [1]

> Section 32–15 of the Zoning Resolution, Use Group 6A, currently includes: Eating or drinking places, including those which provide outdoor table service or incidental musical entertainment [without dancing] either by mechanical device or by not more than three persons playing piano, organ, accordion, guitar, or any string instrument, and those which have accessory drive-through facilities. Such uses are permitted as-of-right in C1, C2, C4, C5, C6, C8, M1, M2 and M3 Districts; a special permit is required in C3 Districts [. . .].[2] Section 32–21 of the Zoning Resolution, Use Group 12A, currently includes: Eating or drinking places, without restrictions on entertainment or dancing. Such uses are permitted as-of-right in C4, C6, C7, C8, and most manufacturing districts and by special permit in C2, C3, M1–5A and M1–5B.[3] Such establishments are not permitted presently in C1 and C5 districts [except that they are permitted in hotels in C5 Districts] (CPC 1989, 2–3; my comment in brackets)

As the above zoning code shows, establishments designated as Use Group 12A were not allowed to locate in some commercial zones, where those designated Use Group 6A were, in fact, allowed to locate (see Figure 2.1). The differential enforcement of regulations between Use Group 12A and 6A, or between cabarets and non-cabarets, was implemented because the CPC understood that the former would invite more crowds and noise, bringing a much more serious impact on the surrounding residential communities than the latter. This rationale has been prone to dispute, and was eventually, in 1986, disputed in court in a case between the municipality and the Musicians' Union. This court case is elaborated upon in Chapter 4.

Under the city's zoning law, each Use Group has three categories applied to the location and the operation of their use. These are 'as-of-right,' 'special permits,' and 'off-limits.' As-of-right development means that a specific use in an establishment, such as social dancing or live music, is automatically allowed in a specific zone that the establishment is located in without it complying with any special requirements as long as the structure that houses the use complies with the Building Code. Special permits were granted to applicants by the Board of Standards and Appeals, which consisted of planners and engineers appointed by the mayor (Chevigny 1991: 10). The conditions through which businesses could acquire special permits were time-consuming, costly and unpredictable, so it was hard for under-financed small businesses to acquire them (see Appendix 1 for the terms of special permits). The zoning resolution, as well as the city's overall land use plan, is drafted by the City Planning Commission (CPC), which is a part of the Department of City Planning (DCP). The CPC creates and assigns land-use zone and uses to each land lot in the city. The zoning resolution, created and modified by the CPC, has the effect of law once the proposal for changes in the resolution is approved—an approval granted, up to 1989, by the Board of Estimate, which consisted of the

Figure 2.1 Zoning configurations of Use Group 12A in 1978. Historical Zoning Maps used with permission of the New York City Department of City Planning. All rights reserved.

city's most influential politicians, such as the mayor, the president of the City Council, the comptroller, and the presidents of the five boroughs (ibid.: 9). The Board of Estimate was eliminated under the new City Charter enacted in November of 1989, and since then, City Council, the main legislative body in the city, has been in charge of approving and vetoing land-use plans (DCP 1990: Preface).

Zoning is one of the legitimate means of state intervention in liberal societies, as was discussed in Chapter 1. Zoning has been a means of re-territorializing the city's economic, political, and social constellations into directions that the municipality deems desirable. For this reason, the CPC works in close collaboration with the Mayor's office. Historically, the CPC has inclined toward mandating special permit provisos rather than downright off-limits provisos to certain land-uses, as the former can give an impression that the municipality does not suppress certain land-uses outright, and therefore preempts possible resistance from affected bodies. It also grants bargaining power to the municipality to extract something from the affected body in return for the permission to run a business (Chevigny 1991: 146).

There is another central apparatus in the land-use planning process in the city. New York City has a unique system that involves local Community Boards (CBs)—neighborhood governing bodies—in land-use decisions (see Appendix 2). Through a process that is now called Uniform Land Use Review Procedures (ULURP), CBs review and vote on almost all land-use change proposals, including development plans, changes in the zoning code and special permit grants that affect their own neighborhoods.[4] They can hold public hearings over proposals for changes in land-use, and make recommendations to the CPC. In 1990, CBs were given authority to prepare plans and submit them to the Planning Commission and City Council for approval (Forman 2000). While these recommendations are non-binding and advisory, this does not mean that CBs play a marginal role in the planning process. On the contrary—and as will be unraveled in the ensuing chapters—the CPC and the Board of Estimate have very much heeded the pleas of CBs, although CBs have often argued that their role in city land-use planning is limited. The Board of Estimate was made up of elected politicians, who naturally tended to be sensitive to popular votes, which explains why it had to be attentive to CBs whose active members are mostly property owners (for similar reasons, City Council has also paid heed to CBs). In addition, since gentrification had become both an important means and objective for the municipality to overcome the fiscal deficit and post-industrializing the city's economy, the inevitable result was that pleas from gentries that have increasingly constituted the majority of active members of CBs have been favorably received by the municipality and City Council. On the other hand, the Borough President and the City Council are closely involved in structuring each CB's acting committee—a structure that furthers the liaison between the city's elected politicians and CBs, and, therefore, helps policy objectives penetrate into, and communicate with, civil society.[5]

Compared to powerful property owners, it has been difficult for politically disenfranchised or socially stigmatized groups to intervene to initiate new zoning rules, or amend existing ones. The uneven power relation inherent in zoning systems has made zoning an arena of acute struggle between different groups in the city. This unevenness is also reflected in zoning regulations that govern nightlife businesses in the city, including the cabaret law's zoning rules. CBs have been institutionally empowered to control the presence and nature of nightlife businesses operating in their neighborhoods. On the other hand, owners of nightlife businesses, especially those that offer entertainments and musicians/dancers/performers related to these businesses, have been primarily considered to be objects that needed management by the CPC and CBs, and have rarely been identified as empowered partners in writing and amending zoning resolutions. This was why Musicians' Union Local 802 in 1986 and dancers in 2005 decided to resolve the cabaret law regulations in court (see Chapters 4 and 8), as it was patently obvious to them that solving problems they had with cabaret law zoning regulations through negotiations with public officials was never going to prove fruitful.

The problems related to the cabaret law did not end with the matter of zoning. The cabaret license application process itself has been (and has been increasingly) cumbersome and often prohibitively expensive. Individual owners start the application process by submitting an Inspection/Recommendation form to the Department of Consumer Affairs (DCA). This form must be submitted with a copy of the Certificate of Occupancy indicating that the venue is located within a zone and a building that allows Use Group 12. This means that if the place is not located in the correct zone for Use Group 12, or has not received a zoning variance that enable the location of businesses of Use Group 12, then the owner of the new establishment cannot even initiate the application process. Submission of the Inspection/Recommendation form typically triggers a host of mandatory inspections. The premises are inspected by the FDNY to check whether the structure complies with the proper fire and safety requirements applicable to cabarets. This part of the inspection procedure can be omitted if the establishment has a current Place of Assembly permit issued by the Department of Buildings (DOB). The premises also need to be inspected by the DCA for compliance with the electrical code. The Community Board (CB) is given forty-five days to submit to the DCA any information about the applicant that they see as relevant.[6] The CB investigates the neighborhood impact that the applying business would cause, and also the past history of the owner's management of other businesses. The CB can bring its findings to the DCA's attention. In principle, the DCA is not obligated to follow the CB's recommendation in its decision whether or not to grant a cabaret license to the business (2–204 of Title 3, The Rules of the City of New York). In reality, however, the reviews submitted by CBs have increasingly been subscribed to as key in the cabaret licensing process.

Once this initial set of inspections is passed, the applicant is instructed by the DCA to a submit a number of documents within sixty days, including an Environmental Control Board Clearance (clearance of any outstanding violations related to subject premises), a Notarized Affidavit of

Security Personnel Background Check and a Cabaret Digital Video Surveillance Cameras Certification (also see http://www.nyc.gov/html/dca/html/licenses/073.shtml).[7] While these documents and the process of submitting them may seem straightforward, cabaret applicants have contended that officers at the DCA processed the cabaret application in confusing ways, and often dragged the process of granting the license out for no transparent reason, as will be seen in the case of 8 B.C. in Chapter 4 and that of Baktun in Chapter 6. Such red tape-related delays in the licensing process has sometimes proved too costly for under-financed clubs, which already have to incur heavy expenses including paying rent, and getting authorized certificates for public assembly, fire safety, soundproofing and the like—expenses that have grown much pricier along with the gentrification of the city. These factors have had the combined effect of discouraging entrepreneurs from opening cabarets, especially those who have limited financial, legal and other relevant resources.

The Department of Environmental Protection (DEP) sets noise rules and inspects compliance separately from the DCA. In late 1972, NYC legislated a noise code that dictated that bars and nightclubs should not emit sound that exceeded forty-five decibels when measured inside a nearby residence. Also, at street level, police officers could cite nightclubs if they determined that the noise from bars and nightclubs consisted of "unreasonable noise" (Hu 2005). This noise code was overhauled in 2005 by Mayor Bloomberg to establish more effective control over various sources of noise including nightlife businesses (see Chapter 7).

While the DCA was in charge of reviewing cabaret applications and granting the cabaret license to applicants, the decision making process of cabaret licensing and especially the enforcement of the cabaret law compliance were diffused among different administrative departments (DCA, DCP, DOB, FDNY, DEP), the Board of Estimate (and later City Council), the Board of Standards and Appeals, CBs and the NYPD. This fragmented process has posed serious obstacles to affected parties (musicians, dancers and businesses owners) in challenging the problems associated with the cabaret law, as we will see in Chapter 7. It was also a hurdle for the municipality in conducting effective enforcement of the cabaret law. This was the context in which the Mayor's office has settled as a body that coordinates these complex processes and fragmentations (Chevigny 1991: 103).

3 Development of Dance Subcultures in the 1970s

An Italian-American, David Mancuso, started to host a party called the Loft in an abandoned factory loft in the NoHo area in 1970, drawing primarily gay and/or black patrons. The Loft was going to effect a sea change in the future of social dancing and nighttime socialization locally, nationally and even globally. This chapter examines the cultural politics formed around social dancing and alternative communities (made up mostly of racial/ethnic minorities) developed in the Loft, and contextualizes it in the city's transition to a post-industrial city. The kind of politics that was created in the Loft and its contemporary parties and clubs took on the character of the "carnivalesque" that Bakhtin (1984) ascribed to popular carnival cultures in the medieval marketplace. Standing in sharp contrast to the medieval ruling-class cultures of hierarchy, authority, orderliness and self-abnegation sponsored and enacted by the Church and the state, the carnivalesque was "temporary liberation from the prevailing truth and from the established order . . . the suspension of all hierarchical rank, privileges, norms and prohibitions" (Bakhtin 1984: 10). It was a Dionysian festival in which the body was celebrated, specifically "grotesque" bodies characterized by and associated with profane words, degradation, disorder, unrestrained sensuality, dancing, singing, over-eating and over-drinking, cross-dressing and laughing—the very opposite of disembodied, cleansed, prohibited and closed official "classic" bodies (ibid.: 24–26). Individual bodies were closely connected to the collective body of people in the carnival, unlike the individualism that later prevailed in bourgeois cultures. The carnival was a site for deviant and undisciplined bodies, and in it, hierarchies and ruling norms were mocked and transgressed, and the world was turned upside-down. NYC's underground parties and dance clubs in the 1960s to 1970s to a significant degree resembled the carnival and its body politics.

However, Bakhtin's conceptualization of the carnival leaves many questions to be answered. Citing Gramsci's more nuanced analysis of common sense and hegemony, McNally (2001: 149) rightly points out that Bakhtin's conceptualization of the carnival failed to see it "as complexes of oppositional *and* dominant values." Alternative cultures are often not radically immune from dominant mainstream values, pose no political threat, are often hierarchical and exclusionary of others, and have been very vulnerable to commodification (ibid.: 149–159). Dance subcultures in this period in the city also evolved into two large axes, with one axis based in the downtown underground, and the other overground, roughly in New York's midtown—although these two sites did intermingle with one another. The midtown dance subcultures were enlisting liminal, transgressive and experimental subcultures—very often the fringe cultures originally cooked up and enjoyed in the downtown party network in a close relation to the particular politics of race and sexuality—into celebrity-oriented hyper-commercialized consumerist spectacle. This chapter examines into what directions the midtown disco phenomena that Studio 54, the mammoth midtown discotheque, was a beacon of, along with the film, *Saturday Night Fever*, have shaped the popular perception of discos and dance cultures.

In this chapter, I also point out that the development of NYC as the center of disco culture broadly coincided with the directive of post-industrialization that the municipality has aimed to achieve since the 1960s. The underground parties and discotheques were often illegally located in abandoned manufacturing zones, but the regulatory atmosphere was quite permissive, as the municipality was in financial doldrums and could not afford regular nightlife inspections. The permissive attitude, however, was also related to that discos and parties often helped to resuscitate buildings in manufacturing and commercial neighborhoods that had been divested in the wake of deindustrialization and suburbanization, and helped to reverse the popular perception of the city as a maelstrom of crisis, crime and dereliction into that of the city as a fun place offering dynamic lifestyles and cultures. In addition, the disco phenomena, as Lawrence (2003) argues, played a part in the initial post-industrial transition of the city's economy, in which service and entertainments industries and consumption oriented urban space increasingly prevail over manufacturing industries and urban space as a site of production.

These parties and discotheques soon started to become the target of community outcry in the rapidly gentrifying SoHo and NoHo areas, and this chapter will provide details about a particular conflict over the re-opening of the Loft in these areas. This conflict presages what would become more common from the 1980s onwards—the "gentrification with and against nightlife."

DIVESTMENT AND ABANDONMENT OF NEIGHBORHOODS IN THE 1960S AND 1970S

During the postwar age NYC took on the general features of cities operating under a Keynesian welfare statism and a Fordist economy; that is, secure jobs with disposable incomes for working-class citizens, strong unions, mass production for mass consumption and federal support for liberal urban government (Freeman 2000). The municipality channeled federal funds into the city's social and physical infrastructure, providing the working class with decent social benefits, affordable and social housing, and affordable or free higher education, although it had become increasingly clear that the non-white working class—blacks and Hispanic immigrants—were systematically excluded from the trickle down effects of the golden age economic boom and the state's liberal policies. The city was also experiencing the federally subsidized suburbanization of working-class households (and later commerce and industries), and the inner-city (especially ethnic/racial minority enclaves) was increasingly confronting the nefarious impacts of divestment (Berman 1988).

By the late 1960s, when the system of Fordism and Keynesian welfare was experiencing falling rates of profits, NYC came into a deepening fiscal deficit. With suburbanization of the white working class and de-industrialization came the decline in tax revenues to the city's coffers. The relocation of manufacturing businesses out of the city undermined the employment base, especially for blacks and immigrants. This resulted in increased demands for public expenditure by the municipality to deliver welfare and social services to these populations (Freeman 2000: 256–57). This condition, along with cuts in federal outlays to cities, ushered in a fiscal squeeze on the municipality, which in turn increasingly resorted to debt financing by getting loans from banks and issuing municipal bonds (Tabb 1982). Tabb (1984: 327) has also argued that an attempt to transform the city's downtown into a corporate center in the 1960s caused the city to borrow money, increasing its debt load. This questionable debt accumulation was encouraged by banks that knew that this would cause the municipality to come into a state of questionable fiscal healthiness, even as they profited from this process, in particular when they raised interest rates in the early 1970s with nefarious impacts on the municipality (Lichten 1986; Tabb 1984). The downturn of the real-estate sector in 1973 that took place as part of the general economic downturn further struck a blow to the city, as it reduced the property tax base.

The worst moment was reached in 1975, when the city's bonds were downgraded; Moody's, for example, downgraded the city credit rating from A to Ba, and then from Ba to Caa (Mitchell and Beckett 2008: 82). This made the municipal bonds virtually unmarketable, eventually bankrupting the city. This opened the door for financial elites to take control of the city's governance and fiscal planning, dispossessing the city of conditions of autonomy (ibid.: 83). A set of (local and non-local) financial institutions such as the Financial Community Liaison Group (FCLG), Municipal Assistance Corporation (MAC) and the Emergency Financial Control Board (EFCB), were created to supposedly "help" the municipality to restructure its fiscal policies. Run by financial elites, these institutions and the policy activism that they promoted, many have observed, was an attempt to restore the power of financial capital from the "excess of democracy" (Tabb 1984: 325) of the Keynesian period (also see Lichten 1986: 108). Despite the complicated and more structural nature of the causation of the 1970s crisis, the financiers that ran these institutions and right wing politicians in alliance vociferously blamed "greedy" municipal unions, "lazy" welfare recipients, corrupt liberal-populist politicians and inefficient and bureaucratic government for causing the crisis. Under this logic, the MAC board put forward a policy regime consisting of punitive austerity on these purported culprits, with cuts in social provisioning (e.g. free or affordable higher education, affordable housing, welfare benefits to the poor) and public downsizing (for example, layoffs of unionized municipal workers), trumpeting this strategy as the only way that the city could achieve fiscal health. It also steered the municipality toward business friendly policies, especially in the areas of volatile, crisis-prone industries like real-estate, producer services, banking and insurance (Mitchell and Beckett 2008: 83). The fiscal crisis period provided a chance for the ruling class to normalize market intervention, especially by financiers, in the governance of NYC.

By the early 1970s, the urban landscape in NYC was littered with dilapidated neighborhoods with a few sporadic islands of middle-upper-class neighborhoods. Many of the dilapidated areas had previously been the habitats of working-class whites but, after this population moved out of the city to the suburbs in a phenomenon called "white flight," these areas were now occupied by working-class blacks and immigrant Latinos. Some of these neighborhoods maintained vibrant communities during the 1950s and the early 1960s. But, as more low-skill manufacturing jobs decamped from the city, these communities faltered. In addition, the municipality's urban renewal programs, proceeding with the objective of "slum-clearance," effected the displacement and eventual disintegration of these communities (Berman 1988; Mele 2000: 145–52). Frustration with these urban conditions was expressed through riots and protests mobilized by minority communities. During this period, neighborhoods previously occupied by manufacturing industries such as SoHo, NoHo, TriBeCa, the Flatiron District and the Midtown Garment District, also experienced gradual abandonment and dilapidation. Even though small-scale manufacturing firms were doing relatively well in the city, the city's financial sector and planners had already planned to de-industrialize the city's economy since as early as the 1950s (Zukin 1989). Through various planning measures, manufacturing spaces had been curtailed and restricted to a few neighborhoods, and planners sought to destroy buildings in these districts to establish office complexes and housing quarters there. These plans of destruction were thwarted in places like SoHo and NoHo by the middle-class-led, often Jane Jacobs-inspired, movements of preservation of historic urban architecture often in alliance with artist groups.

During the city's fiscal crisis, disinvestment in the city's economy and real-estate was further fueled.[1] With the virtual paralysis of essential city services, the impoverishment of the working class in the disinvested neighborhoods deepened further. The dominant discourse about the city was one of despair, culminating in the highly publicized image of the looting mayhem during the blackout of 1977 and the "Son of Sam" slayings. By the late 1970s, however, signs of property market revival in some neighborhoods emerged together with a recovery of the city's economy, which I continue to examine in the next chapter. In particular, the Upper East Side was rapidly gentrified, and SoHo and NoHo dramatically gentrified through growth as an art district. Poor artists who illegally squatted in lofts in these two neighborhoods in the 1960s were spared penalties as the municipality legalized artists' residences by amending the zoning designation of the neighborhoods from M1–5 (manufacturing zones) to M1–5B and M1–5A in 1971.[2] As the real-estate market's interests in the neighborhood increased, and more and more non-artists started to live there, the municipality revised the J-51 program in 1975 to provide tax benefits for large-scale residential conversion of lofts. This was followed in 1976 with the legalization of artists' residential conversions in NoHo, and also by the creation of a neighborhood called TriBeCA, in which residential conversion of lofts for artists and non-artists was legalized (Zukin 1989: 53). SoHo in particular would set a landmark example in the future of modern urban planning, by showing how arts and aesthetic ambiance created by them in a neighborhood could become important triggers of gentrification, and how poor artists could act as a "critical infrastructure" for this type of transformation of a neighborhood (eventually to the disservice of their own presence in these neighborhoods).

DEVELOPMENT OF DANCE SUBCULTURES IN THE EARLY TO MID 1970s

Contemporary historians and ethnomusicologists regard the Loft, a regularly held party conceived by David Mancuso in 1970, as the origin of a dance subculture later named disco (Berry 1992; Fikentscher 2000; Lawrence 2003). Here, drawing upon Tim Lawrence's (2003) extensive ethnographic work on dance subcultures of NYC in the 1970s, I offer a condensed version of this history. Mancuso was influenced by the 1960s East Village venues that were characterized by an unconventional and psychedelic philosophy and ambiance of intimacy. As Greenwich Village gentrified in the 1950s to 1960s, many progressive joints that housed hippie cultures moved to the areas that are now called East Village. These venues included Electronic Circus, which opened in 1967 and featured Andy Warhol and Velvet Underground, and the Fillmore East, a hippie venue that opened in 1968. Both had live entertainment, and in the words of Mancuso, were "mixed, very integrated, very intense, very free, very positive" (Lawrence 2003: 8). Mancuso also frequented "rent parties" that were hosted in private apartments and that had a homelike feel, and were sites of experimentation for multiple cultural practices (ibid.).

Mancuso started to have his own weekly rent party in 1970 at 647 Broadway in a neighborhood that would be later called NoHo. Like SoHo, at the time NoHo was a deserted manufacturing district with disused buildings in which poor artists, underground dance partiers and experimental jazz musicians, were illegally living, and often setting up creative networks among each other. Mancuso's parties, popularly called the Loft, operated as a venue in which living, partying and arts were fused together. The Loft became established as a key player within the bustling artist communities in SoHo and NoHo (ibid.: 9). Mancuso decorated the ceiling of the dance floor with balloons of various colors together with a huge mirror ball in order to increase the festive atmosphere of the party. Liquor was neither served nor sold on the premises. The party was also private, where only invitees could get in. Many of the attendees were black, Latinos, Italian Americans and/or gay, or came from other economically and socially marginalized groups.[3] For these contingents, the Loft was "their own little artificial paradise" (ibid.: 53) in which they created a tribal community and family among themselves that were different from what mainstream society defined, and that provided bodily and emotional release and a temporary escape from the economic and social distress that they were confronted with in the urban milieus in which they were living (Fikentscher 2000). The bonds established between participants were based on the shared feeling and ecstasy communicated through collective dance, music and LSD.

"Hardcore dancing" was the most important activity in this party (Lawrence 2003: 10). The dancing was interactive and collective rather than individual or partnered. There was an impassioned interaction between the DJ (Mancuso) and the dancers; the DJ always improvised in response to the way the dancers were responding to the music. Contrary to partnered dancing prevalent in the previous years, dancers in the Loft danced individually; yet individualism was not the organizing code in the Loft. Instead, individuals communicated through music carefully orchestrated by the DJ and the collective energy on the floor, elevating the rapport formed on the dance floor into a ritual. The format of intimate interaction between the DJ and the dancers, the atmosphere created on the dance floor through interactive processes and with the energy, style and enthusiasm of the dancers, and the organic community premised on shared appreciation of music and dance became the standard feature of underground dance cultures. The Loft also commenced a new way of socialization and community building within nightlife (ibid.: 1–3). The future DJs that would dominate downtown Manhattan—for example, Francis Grasso in the Sanctuary, who would create a new concept of DJing by introducing the "mixing" technique, Nicky Siano in the Gallery and Larry Levan in Paradise Garage (the last of which will be introduced in Chapter 5), and many others—were regulars at the Loft.

For Mancuso, the ethos of the Loft party was similar to that which had emanated from the countercultural movement in the 1960s. He said in one interview:

> I was on the streets and in the party. Dancing and politics were on the same wavelength, and the Loft created a little social progress in tune with the times. (Lawrence 2003: 51)

That is, like in the hippie movements, the Loft created a condition in which established norms were turned upside down, bodily liberty was celebrated and stigmas against sexual/racial/gender/class identities of the social margins were cancelled out, however temporary and seemingly apolitical these parties might have been on the surface (Lawrence 2006: 129; also see Fikentscher 2000).

From the early 1960s, as the legal restraints on homosexuality were gradually loosened, gays more actively pursued the freedom to have fun in public (Lawrence 2003: 29–30). As was exposed in the Stonewall riot in 1969, dance venues were one of the sites where homosexuals socialized. In the 1960s, the area around Sheridan Square in the West Village and Fire Island in Long Island had several places where these contingents hung out. But the Sanctuary, an underground gay discotheque located in the derelict working-class neighborhood of Hell's Kitchen, was the most prominent gay club, whose atmosphere took after that of the Loft and which had the resident DJ Francis Grasso developing a dexterous form of disc mixing. At the same time, in Harlem, the competitive ballroom walk parties of black drag queens—of the kind that would be featured in Jennie Livingston's 1990 film, *Paris is Burning*—started to earn a wider reputation (ibid.: 46). The gay nightlife scene received a further impetus in 1971, when the administration of Mayor John Lindsay relaxed rules that banned men dancing with each other (Narvaez 1971). More venues opened up afterwards, and DJs spinning in gay venues started to acquire a citywide reputation.

It also became increasingly common that clubs caused friction whenever they were located in proximity to residential communities. Sanctuary's wild and boisterous revelers invited frequent complaints from the working poor in the neighborhood, and in 1972, the club was closed after a citation for public nuisance and widespread drug use within the club (Lawrence 2003: 66–67). Tambourine and Superstar in the Upper East Side, which was experiencing gentrification at the time, were also closed down for reasons related to violence and drugs (ibid.: 68, 93). Nevertheless, discotheques continued to open more quickly than they closed (ibid.: 96). These new venues took after the Loft's party template—e.g. the invitation system, the sound system that Mancuso originally invented, and dynamic DJ-dancer interaction. The Gallery, which opened in 1973 in a loft in the Flatiron District as an illegal underground club, was the most prominent example. Nicky Siano, the resident DJ of the club who used to frequent the Loft, aimed to create a heterosexual version of the Loft (even though many of the Gallery's patrons were also gays), and became another powerful ruler in the city's underground night scene.

In late 1973, Vince Aletti wrote in *Rolling Stone* that the discotheques that originally belonged to the hardcore dance crowd—blacks, Lationos, Italian Americans, gays—had now become "a rapidly spreading social phenomenon (via juice bars, after-hours clubs, private lofts open on weekends to members only, floating groups of party-givers who take over the ballrooms of old hotels from midnight to dawn)" and also "a strong influence on the music people listen to and buy" (cited in Lawrence 2003: 115–16). According to Lawrence (2003: 116), Aletti's piece was a sign that clandestine dance parties and clubs that had been amorphously developing in the city's derelict corners had become by then a "recognizable scene" with significant public impact. In addition, contrary to the unorganized and scattered features of its initial period, the scene now acquired an organism-like momentum, through which nodes that constituted the "downtown party network," such as clubs, record stores, parties, radio stations and (independent) record companies, interacted dynamically with tremendous stylistic developments in dance music and cross-fertilization among different genres. To fight the uneven development among over- and under-ground clubs precipitated by the infiltration of the major music industry, David Mancuso created a Record Pool in 1975, an organization of the city's DJs (ibid.: 158). The pool sought to promote the democratic redistribution of the records acquired from record companies, and other issues that affected DJs, parties and discotheques in the city were discussed among the pool members and democratically decided upon by voting among members.

CONFLICTS IN CHANGING NOHO

In 1972, the Loft was raided by the police after receiving noise complaints from the neighbors upstairs. The police questioned the safety of the venue, but Mancuso demonstrated to the police that he had a perfect sprinkler system in place and emergency lights, so the police did not bother Mancuso's party after their initial inspection (Aletti 1975: 127)—a common approach by the police in 1970s to illegal but harmless venues. However, the Loft became an increasing target of neighbors' complaints in 1973 when the Broadway Central Hotel, located a couple of blocks away from the Loft, collapsed. Loft was located in a similarly old building, which made it subject to a similar collapse. Concerned neighbors argued that Mancuso's modification of the interior of the Loft venue would undermine the safety of the building due to the vibrations from Mancuso's enormous sound system and the stomping of "500 frenzied dancers," from what they called "some of the more bizarre segments of our society" (Thomas Jr. 1974: 45). Neighbors further argued that Mancuso had created a firetrap with the soundproofing equipment, had caused neighbors inconvenience by frequently changing entrance keys, shattered their serenity, and took over and ruined what they regarded as their studios and their homes.[4] These complaints from neighbors ushered in stricter inspections of the Loft by enforcement agencies, and because the venue was missing important legal documents (including basic administrative paperwork, including a Certificate of Occupancy and Public Assembly Permits), Mancuso's argument that the party was safe did not hold much water. The Department of Buildings considered the situation to be too sensitive since neighbors were increasingly panic-stricken, and ordered Mancuso to vacate in June 1974 over charges of "illegal openings" in the wall and "overcrowding" (Lawrence 2003: 126).

The Broadway Central Hotel collapse also drew attention to other venues that were operating in former manufacturing-use lofts. Nicky Siano's Gallery was also closed down. After the closing of the Loft and the Gallery, the SoHo Place opened in 1974 and became the leading loft party in town. As Richard Long, the owner of the venue and the best sound expert in the city at that time, started to use the place as the showcase for the sound systems that he was building, the party started to gain the notoriety among neighboring residents. But, one of the staff members of the party also recounted the conflicts in terms of the neighbors' racism, saying "It was also unheard of to have blacks living in SoHo at the time" (cited in Lawrence 2003: 134).

DISCOS AND POST-INDUSTRIALIZATION OF THE CITY

The disco became a national phenomenon by 1975. The coverage of disco was not confined to media specializing in music, such as *Billboard. Newsweek* and *Wall Street Journal* also analyzed the giant economy of the disco industry, and lifestyle magazines got busy introducing styles associated with disco culture. NYC was at the center of the action as far as disco was concerned, and discos emerged as one of the few vital industries of the city during the economic crisis of the mid-1970s. More and more young populations—mostly from NYC's suburbs—spent their weekend nights in the city's dance spots. According to the Department of Consumer Affairs (DCA), forty-five new cabaret licenses were issued in 1975 only, and this did not include already licensed clubs that had started to include discotheque dancing (Kennedy 1976). Everyone wanted to go to the disco, and even pizza chains installed turntables and called themselves discotheques (Weller 1975).

According to Lawrence (2003: 181), New York's growth as the center of disco was consistent with the era's "profound shift from the production of goods to the production of events in the American economy of the 1970s." While Lawrence does not elaborate on this point further, he seems to mean that discotheques were one of the signifiers, and at the same time, constituents, of the post-industrial transition of the city's economy, in which production of high-end service and consumption of entertainments and culture (and the associated labor force) were prioritized over the production of tangible manufacturing goods (and its related labor force). Under the supervision of institutions like MAC, the municipality was forced to neoliberalize and post-industrialize the city's economic basis, and was anxious to transform the image of the city, which had until recently been associated with crime, homelessness and riots, into a brand that was appealing to investors, corporations and their workforces (Greenberg 2008). The vibrant nightlife to which the disco boom contributed, and the peculiar lifestyle consumption that it seemed to promise, were therefore placed to realign the popular imagery of NYC, especially to the baby boom generation suburbanites for whom diverse lifestyle options of their habitats was becoming increasingly valued (cf. Ley 1996). Discos were also counted as an important resource of tourism, one of the most prominent post-industrial businesses.

Coincidently, many discotheques and parties—one of the prominent symbols of the new urban economy of "events"—were located in the former manufacturing districts or run-down immigrant working-class neighborhoods, literally replacing industrial landscapes with their own, and thus further adding to the sense of post-industrialization in the city. Initially, artists or partiers chose to locate in these abandoned sections of the city primarily to cash in on the cheap rents and large loft spaces, a phenomenon that converged with the interests of landlords that could not find other tenants for their derelict buildings. But, by 1975, having an address in such neighborhoods had become the "thing" to do for chic discotheques as well as trend-setting galleries. For example, even the upscale disco, Infinity, strategically chose its location in a factory space on lower Broadway, trying to take on the "feel of mystique and secrecy" that had hitherto been exclusive to downtown underground clubs (Lawrence 2003: 187). Accommodating such post-industrial businesses as discotheques in the declining industrial districts would not only raise the value of the derelict neighborhood and the built environment in it, but also ultimately helped to normalize the eviction of manufacturing tenants and de-industrialization in general.

Discotheques also emerged as an icon of a democratic melting pot that urban life was all about. In the media, discotheques were trumpeted as "egalitarian," and as "[transcending] social, ethnic, economic and even sexual lines," compared to other entertainment in the city (Kennedy 1976: 29). This aspect of discotheques must have had certain "sign values" to baby boom generations who grew up in the heated milieu of radical identity politics, as well as mass consumption, and had developed tastes for differences and aesthetic cosmopolitanism as one of their key kinds of "cultural capital." The irony was that the lines of racial, ethnic, sexual and economic differences may have dissolved on these dance floors, but this was only done stylistically. On the dance floor, white and/or heterosexual dancers accepted and imitated the styles associated with the subcultures developed among ethnic, racial and sexual minority communities, but it is not clear how this stylistic miscegenation may have actually promoted democracy in the real world. Some discotheques strategically placed black gay contingents on the dance floor to make the ambiance of the venue look merrier and even "authentic" (Lawrence 2003: 178) This was the manifestation of the commercial version of the "carnivalesque" and the transgression and liminality of body politics as these were placed under calculus of profitability. Despite the cosmopolitan orientation of the images promoted by clubs and their marketing schemes, some discotheques patronized by suburbanites and tourists indeed made efforts to discourage elements that would make the venue look like "fag joints" (ibid.: 185).

The cosmetic egalitarianism of discos was also demonstrated in the irony that the "democratic" conflation of multiple styles (and often multi-ethnic/racial mixing) on the dance floor emerged simultaneously with the growing economic and social marginalization that the majority of original architects of disco—the poor working class of black, Italian Americans and/or gays—were going through under the state's turn toward neoliberalization and post-industrialization. The parties and the styles that these populations on the margin of the society had developed as a "ritual" of celebrating and negotiating their marginality were transformed into a "spectacle" of elusive egalitarianism and a post-industrial version of multiculturalism that would effectively paper over the accelerating marginalization of these persons.

THE NEW SOHO AND NOHO

In the midst of the disco boom that swept the nation and the city, on the neighborhood scale, disco parties and clubs got into trouble. This was particularly true of the SoHo and NoHo area to which the booming disco culture was native. SoHo Place, finally closed down after Long was taken to court due to complaints by neighbors over noise generated by the club. Mancuso, having been forced to close the Loft at Broadway in the NoHo neighborhood, tried to open the same party in an almost vacant block at the northern edge of SoHo, but immediately came up against opposition by residents from nearby. Opening a party at that address was a daunting project by then, as SoHo was evolving into an art district combined with upscale residences. *SoHo Weekly News* (*SWN*)'s editor led the campaign against the Loft, claiming that it was an "after hours, drug-oriented club" (cited in Aletti 1975: 127).[5] The *SWN*'s approach mirrored, and further stirred up, SoHo residents' fear about negative impacts that some businesses would make on the emerging artist identity of their neighborhood. Regarding the re-opening of the Loft, the SoHo Artists Association announced, "This is the beginning of an invasion," and anxious housewives rallied with pickets saying "No Juice Joint in SoHo" (ibid.). The opposition was sometimes articulated in racist terms. One *SWN* staffer recalled that the *SWN* editor's notion of the Loft was that of "a drug haven, with people out in the street, muggings, blacks" (ibid.). In another case that involved the (re)opening of the Gallery in the neighborhood, the *SWN* opined that the club would have a predominantly black crowd that would make lots of noise, whereas as to the opening of Flamingo, on the other hand, it commented that the patrons would be white and very discreet (ibid.).

In response to the community rally, Mancuso announced a "Declaration of Intent" in which he argued that the Loft had been law-abiding, had exerted good values (of dance and fellowship) that people had lost in their growing years, and had given employment to deserving young people (ibid.: 127). To avoid any doubt about whether his party was operating legally or not, Mancuso applied for, and acquired, a Certificate of Occupancy and Public Assembly Permits. Realizing that this might not be enough, however, he began the application for a cabaret license. The hearing for this application was long and marked by frays between SoHo residents and Mancuso's party crew. But, it turned out that the Loft did not even need to apply for a cabaret license, because it was an invitation-only, and thus private, party, and did not sell alcohol. In these cases, clubs did not require cabaret licenses. This was an important discovery for the city's underground parties and clubs, and from this point on, became the legal precedent of party operations. The Loft, after having gone through these conflict-ridden processes, officially reopened in October 1975.

In his interview with Aletti, Mancuso said that he was not an invader in the community in which he wanted to newly place the Loft, as he had himself lived just outside SoHo for the past eight years. He lamented that the area was not for people with new ideas anymore, and that there were no more struggling young artists. Instead, it had become a place for "artists turned high bourgeoisie," the "troops of the chic and the super-chic," "the rich who are concerned about dog leashes and the crime rate" (ibid.: 127). While the Loft was able to escape being shut down, the battles over parties and nightlife businesses in SoHo and NoHo continued, as the gentrification of these neighborhoods invited more commercial entertainment businesses into the neighborhood (Schuman 1974). These battles culminated in 1976 in the zoning modification of SoHo and NoHo that would ban the opening of any new eating or drinking place of more than 5,000 square feet, as well as theaters, clubs, banquet hall, caterers, public dance halls, cabarets and large sports facilities (interview, Ken Bergen, CPC staff planner, personal communication, 11 March 2005). Restaurants with less than 5,000 square feet and theaters with fewer than 100 seats would still be permitted. Reporting that within the last year, four new discotheques had opened in a five block stretch of SoHo, the City Planning Commission (CPC) at the time, said:

> This type of use exploits the energy of the art center but contributes nothing to its growth . . . In fact, it threatens the continued viability of the area by making it a less desirable place to work and to live. (Fowler 1976)

This move by the CPC was made after a public hearing in July 1976, where several artists' groups and non-artist residents in SoHo came out to call for a ban on "noisy and offensive" discotheques. The only Commissioner planner dissenting to the zoning change in the CPC opined that while he agreed that discos had been a nuisance for the community, the move would ban establishments that "give the area its special character—innovative, different and creative" (cited in ibid.).

By 1976, SoHo and NoHo had evolved into desirable middle-upper-class habitat with the identities as art districts, and zoning modifications like the above were one type of measure taken to reinforce this emerging identity. The artistic ambiance established by young, obscure artists in these neighborhoods in the 1960s and early 1970s revalorized the up-to-then depreciated property market there, and later attracted real-estate capital which converted lofts into up-market residences. While non-artists were already living in lofts in SoHo with the acquiescence of landlords, the local building codes and zoning restrictions still outlawed loft living for non-artists, and this inhibited banking institutions from loaning out money for loft development (Zukin 1989: 10–11). In 1975, the legal codes lying in the way of developing residential lofts in these neighborhoods were eliminated, and section J-51–2.5 of the Administrative Code of the City of New York was amended to offer long-term tax abatement and tax exemption to developers and owners who would undertake the residential conversion of large commercial and manufacturing buildings—an achievement that was the result of years of lobbying by real-estate developers (ibid.: 13). This paved the way for professional real-estate developers to enter into the loft market, and banks started to finance these developers more willingly. This led to rapid increases in housing prices, and soon most of the resident artists in these neighborhoods could no longer afford the rents. Later, a change in the city's subsidy program for artists dealt the final blow to artists. The municipality withdrew its housing subsidy for artists, converting them into subsidies for art production. This change left artists with no legitimate claim to SoHo as their habitat, unless they could afford to buy a space in this area. By the end of the 1970s and early 1980s, art production almost disappeared from SoHo, and was replaced with upscale boutiques, galleries, and lofts for affluent artists and non-artists.

This process exemplified the vulnerability of the arts to exploitation in the real-estate market, despite the apparent "agency" of arts in engendering positive use values of our derelict built environments. It is also worth noting here that the displacement of artists in these neighborhoods was actually preceded by that of nightlife. Nightlife businesses, including underground parties and clubs, were objected to by an alliance of middle-class transplants and artists' organizations that were seeking to establish the areas' identities as artistic mixed-use communities with a reformist middle-class character, and successfully pressured the CPC to make zoning modifications as such in 1976. This process revealed that nightlife was mainly identified as nuisance "commerce," but hardly as cultural institutions. The process also exemplifies how different types of arts and cultural commerce are valorized differently in the "culturalization of gentrification" and even pitched against one another in their struggles over urban space. In this struggle, the wilder ilk pave the way for those who conform to the taste economy of gentries and the quality of life of a gentrifying locale. This type of sanitization further served to ease private capital's entry into the area, and contributed to creating more suitable conditions for upscale real-estate markets, which eventually priced out artist groups themselves.

COMMERCIAL MIDTOWN DISCOS

By 1977, disco had exploded in popular culture and in nightlife landscapes, incurring exponential profits in the music industry, accounting for almost 40 percent of Top 100 hits, 20,000 discotheques in the country, 200 all-disco radio stations and 8,000 professional DJs (Cheren and Rotello 2000: 242). As *Newsweek* reported in 1978, "[w]ith insidious speed, the after-hours music first heard only by small urban groups of blacks, Hispanics, gays and insomniacs had invaded the hearts, minds and feet of all ages and classes" (cited in ibid.: 243). Two icons that would define the 1970s disco climbed to prominence in 1977: the first was Studio 54 which opened in the midtown of Manhattan in the heart of the Broadway theater district, and the second was a film titled *Saturday Night Fever* that featured the Brooklyn discoscape. Both elevated New York disco culture as a pop culture phenomenon that defined the decade, but they also contributed to reinforcing a "decontextualized" and "dehistoricized" representation of disco (Lawrence 2003: 307). *Saturday Night Fever* represented disco as being related to a hyper-heterosexuality and the seductions of couple-dancing (ibid.: 130), in opposition to the original disco which was characterized more by polymorphous sexuality, a fantastic aural experience and communal celebration of body and identity.

Studio 54 was a mega discotheque that focused on spectacular drama maximized through multimedia sets, the presence of celebrities, discriminatory door policies, a blatant drug culture, musical predictability, hyper-commercialism and disengagement from dancing and listening. Taking after Studio 54's characteristics, a cluster of midtown discotheques emerged in this period (ibid.: 333), forming one axis of two contrasting nodal points of the disco topography in the city:

> One is low-key and secretive, while the other is brazen and publicity-hungry. One embodies downtown, while the other symbolizes midtown. One is focused on the DJ and the dance, while the other is more interested in flashing lights and spectacular sets. (ibid.: 3)

These two scenes of disco, that is, downtown/underground and midtown/overground, were, as Lawrence (2003) argued, interlaced with one another. Studio 54 imported some of the concepts, visuals, and music that were experimented in the Loft and Gallery, and underground DJs also worked in midtown discotheques. The lines dividing these two scenes may have often blurred and been ambiguous, but they evolved, roughly, according to different historical trajectories, objectives and philosophies. In 1979 when the disco craze closely associated with the midtown spectacle ended with the closure of Studio 54, and the "Disco Sucks" chant in Chicago stadium led by the rock DJ Steve Dahl, the downtown underground dance rituals were still going on. The Loft still had its regular Saturday party, and even though some underground parties were closed, including the Gallery (due to a drug overdose by resident DJ Nicky Siano), a new club called Paradise Garage opened in 1977 on King Street in SoHo that would dominate the New York underground scene in the 1980s.

Despite the durability of underground, and more organic, dance communities, popular and academic understandings of dance cultures of the city were limited to their commercialized version. For example, the music critic Albert Goldman criticized the disco culture in *Esquire* as "the final rape of the human sensorium" and the culmination of "narcissist self-indulgence" (Lawrence 2003: 318). When disco became severely criticized by the mainstream commentators and also by leftists as the icon of degenerate commercialism and the hedonistic extreme of capitalism, one lone leftist, Richard Dyer (1979), came out to defend disco.

DISCOS AND CITY MARKETING

In October, 1976, a fire broke out at the Puerto Rican Social Club in the Bronx that killed twenty-five people. This was one of the social clubs popular among ethnic minority enclaves (especially among Hispanics') in this period that operated as illegal after-hours clubs. These clubs provided entertainment like music, dancing, and drinking, and use of marijuana and cocaine was prevalent. These illegal clubs were well known to law enforcement officials, and there was a demand for stepping up inspections as these were also conceived as sanctuaries for prostitutes and petty criminals; little, however, had been done to rein in them. An *NYT* article (Narvaez 1976) reported that disputes between the State Liquor Authority (SLA) and the NYPD about the jurisdiction responsible for regulating these illegal clubs delayed the effective inspections of these clubs. The other reason why these activities were not contained was also because there was an approach in the government to look upon these clubs as "a social necessity and a means of keeping roots in shaky communities," that were often "the only signs of life in otherwise deserted and burned out sections of the city" (ibid.: 1, 46). The article further delivered voices of owners of legal clubs and bars who argued that it was unfair that illegal clubs were operating without paying expenses that would otherwise be incurred to get permits, as well as taxes. In response, the Deputy Mayor said that due to budget constraints, the police would not be able to control illegal clubs and, instead, would work on other issues of greater immediate public safety concern. Two DJs who worked in this period also confirmed to me that in this period, controlling illegal clubs was less of a priority to the enforcement agencies, due to the crime and violence that the city was fraught with in the midst of neighborhood dereliction in the city (interview, Danny Krivit, who DJs at Body & Soul and The 718 Sessions, Dec 22, 2009; interview, Richard Vasquez, a former owner of underground dance club and DJ, personal communication, August 23, 2006).

Apart from a sympathetic approach owing to a recognition of the social importance of nightlife and budget constraints on inspections, the municipality seemed to welcome discos' marketing power. In 1978 *Billboard* hosted Disco Forum IV in NYC, for which Mayor Koch sent a welcoming endorsement:

> The Disco and its lifestyle have helped to contribute to a more harmonious fellowship toward all creeds and races . . . The beat of the disco is the heartbeat of what makes New York City a special kind of place—vitality, individuality and creativity. People all across America made the disco popular—New York City made it a phenomenon. (Mixmaster 1978, cited in Lawrence 2003: 308)

After this announcement, Mayor Koch proclaimed that the week of the Disco Forum would be an official "Disco Week" in the city (ibid: 308).

Such an endorsement may be reflective of a growing perception of the value of nightlife in the post-industrialization of the city's economy, as was previously mentioned. Even though the scene of Broadway theaters weighed more heavily among marketing symbols than other types of nightlife, the city's marketing strategy often spotlighted the city's flourishing nightlife, as it sought to evoke potential post-industrial symbols, such as a convivial urban lifestyle, that were appealing to tourists and professional workers (Fleetwood 1979). Occasionally, the city's promotional brochures featured how nightlife was providing an aura of vitality and as a cultural vanguard to the city (Greenberg 2008: 207–11). The type of nightlife highlighted in the promotional brochures included sophisticated dining and genteel live music experiences. But, the wilder part of nightlife, being niftily and trendily repackaged, took part in this promotion as a positive icon, as a sign of the presence of a variety of interesting cultures in the city. In one promotional campaign in 1978, Studio 54 was featured with the disco song "I ♥ NY" and a message that "hundreds of colorful and interesting people [keep] New York alive at night" (ibid.: 210).

The real-estate sector also followed the suit of disco fever. Horsley (1978) reported that the recent success of big discotheques and the movie, *Saturday*

Night Fever, contributed to the rush by developers to rent out spaces for discotheques. Discotheques were viewed as one of the few economic sources remaining in derelict neighborhoods, as they reduced vacancies, effected face-lifts of crumbling buildings and incurred revenues for landlords and real-estate professionals by occupying properties that were otherwise unleasable for other uses, until the time of reinvestment. This was one of the reasons why the municipality was tolerating discotheques and parties, despite their perceived criminality. The developers found this profit projection particularly apt in the midtown Times Square area, where in derelict buildings sex related shops were main economic resources. As an example, the former Henry Miller Theater on West 43rd Street, for many years a movie theater that showed pornographic films, was converted in the late 1970s into a disco called Xenon. The conversion of abandoned buildings to discos in Times Square was touted as a sign of the revival of Times Square, which was expected to bring big corporations' headquarters back to the area along with hotels and shopping malls (Fowler 1979; Oelsner 1978).

These sympathetic and hospitable gestures to discos and dance subcultures show the complicated and contradictory history of nightlife in the city. Discos and other nightlife businesses and parties continued to contribute to the city's post-industrialization and gentrification in manners that, from the early 1980s onwards, would only fuel an anti-nightlife rallying cry that would seek to crackdown on these businesses and protect quality of life in gentrified space from them. I elaborate this process in Chapter 4, with the concept of "gentrification with and against nightlife."

EAST VILLAGE AND THE PUNK MOVEMENT

While SoHo, NoHo, TriBeCa and Times Square were evolving into a site for dance subcultures, rock and punk subcultures were being incubated among the middle-class youth that sneaked into abandoned buildings in the East Village (as the Lower East Side later came to be known). The East Village was one of the places in NYC that experienced severe disinvestment from the 1960s to the 1970s, especially the far eastern part called Alphabet City, where Puerto Rican immigrants had settled, and the Bowery district, which was perceived as NYC's "Skid Row." The rock and punk subcultures here developed as a social and stylistic critique of, and often escapist and nihilistic outcry against, the unbridled unemployment and impoverishment that the material conditions of this neighborhood represented (Mele 2000: 213–214). This explains why the punk developed here was characterized by "symbolic violence and aggression," which outsiders perceived as "antagonistic, antisocial and threatening" (ibid.: 214). These subcultures produced "glam" or "glitter rock" and later "New Wave," and by the mid 1970s, New York Punk came of age with the emergence of such groups as the Velvet Underground, Iggy Pop and the Stooges and the New York Dolls.

The existence of this youth subculture breathed into the neighborhood a strong countercultural identity in the 1970s and early 1980s. The East Village subcultural scene constituted one node of the "downtown scene," along with other subcultures developed in other parts of downtown, including dance subcultures (ibid.: 215–19). By the end of the 1970s, these multiple nodes of the downtown scene shared a rejection of the midtown scene (especially the mega disco scene) that symbolized extravagant, status-oriented, hyper-commercial and sugar-coated vulgar mainstream cultures. As Lawrence mentioned (cited above), downtown took on a meaning that not only referred to the geographical location, but also an aesthetic or genre opposed to the conventional, corporate, homogenous cultural life. It was also in the 1970s that in the East Village, "club culture" became prominent. "Clubs" became the focal point of the experimentation of arts, music, dance, performance and fashion (with the liberal acceptance of drug and sex), but it was also the site where new styles and cultures were disseminated and consumed by a wider middle-class audience (ibid.: 217). CBGB on the Bowery was well-known in this regard. The popularity of the East Village counter-cultural scene among the middle class would soon turn into the neighborhood into a site of interests in the real-estate market in the 1980s. This will be discussed further in the next chapter.

CONCLUSION

The Loft party, set up in a derelict loft in NoHo, developed a ritual in which participants played a particular expressive body politics through dancing, and in doing so, collectively celebrated their social, racial and sexual marginality. The history of the Loft and other underground parties points us to the values of social dancing: it is a medium for a group of people to express and communicate in a non-verbal, interactive way, and in doing so, to form an alternative community with an alternative philosophy of sexuality and race. This is a key lesson that this chapter delivers that I will revisit toward the end of this book. Another major point of this chapter is the standing that discos have had over the course of the post-industrialization of the city's economy and space. They became one of the elements that catalyzed the transition of the bases of the urban economy "from the production of goods to the production of events." Not only were discos major tourist destinations, but they also helped to reinforce the image of the city as being vibrant, dynamic, fun and tolerant to diversity. At the neighborhood scale, the presence of raving nightlife as well as artistic communities (as was the case in SoHo and NoHo) could stylize the images of these neighborhoods and attract the attentions of potential gentries. Discotheques and parties were also located in derelict neighborhoods, most of which were in increasingly obsolete manufacturing districts, literally facilitating de-industrialization by their presence.

Discotheques and parties came into conflict with neighboring residents, especially (if not exclusively) in gentrifying neighborhoods, due to their nuisance effects and often the racial constitution of their patrons. They were blamed as anathemas to public safety and to the (potential) residential character of neighborhoods. SoHo and NoHo were neighborhoods of just this sort, but they had a more ironic aspect. Here, artists were part of the broader neighborhood alliance that protested the location of nightlife businesses in the neighborhoods. But the efforts of these artists to sanitize the area from commerce associated with public nuisances eventually backfired, bringing about their own displacement by inviting in the upscale real-estate market. Such ironies would reappear from the 1980s on, when more abandoned parts of the city—also neighborhoods of alternative subcultures—began to be gentrified, throwing up struggles over space and culture among multiple actors. I examine this phenomenon in the next chapter.

4 Gentrification with and against Nightlife

1979–1988

When Mayor Ed Koch took office in 1978, the municipality was facing an inevitable dilemma with the city's prospering nightlife: how to balance the flourishing nightlife, which contributed to enhancing the city's image and to boosting its tourism revenues and real-estate values, with the snowballing rallying cries from gentries clamored for the enhancement of quality of life. The political choice was to tighten law enforcement and possibly change regulative rules so as to keep the clubland of the city in line with the quality of life demands of upper-middle class residential property owners. This chapter elaborates on this process of "gentrification with and against nightlife" (see also Hae 2011b). The detailing of this process will demonstrate that gentrification is a complicated process with contingent and contradictory turns that are shaped by multi-faceted struggles over what counts as legitimate urban space, culture and values.

In this chapter, I further bring to attention important points that explain why conflicts over nightlife developed the way they did in the 1980s. First, dance clubs had been perceived by the city's residential communities and the city's enforcement agencies as the number one enemy of the public interest in gentrifying neighborhoods. Cabaret law regulations, as well as other rules (such as safety regulations), were tightened drastically in order to better regulate club businesses, such that it became much more difficult to acquire a cabaret license for clubs located in zones that were not as-of-right for cabarets. As the same stringent cabaret law zoning and licensing rules were applied across the board irrespective of clubs' sizes, this enforcement added financial difficulty to under-financed small clubs that were already struggling with rising rental prices, in particular live music clubs. This was why musicians finally decided to take the cabaret law to court to challenge its constitutionality.

Second, I examine how a First Amendment right can be effectively used as a critical resource by groups associated with specific expressive cultures to challenge anti-nightlife governance. In investigating how groups with certain cultural concerns are able to take advantage of First Amendment rights, I revisit the idea of the "politics of rights" (Scheingold 1974) discussed in the Chapter 1 of this book. I illustrate the potential of First Amendment challenges by certain cultural groups through examining the litigation which the Musicians' Union, Local 802, proceeded with in 1986 and 1988 (*Chiasson v. New York City*). This litigation challenged the constitutionality of the cabaret law's discrimination against live jazz music, specifically the discrimination that regulations against the number of musicians and the types of instruments musicians played represented. The details of the litigation and the court decision of *Chiasson* provide important contexts for understanding the City Planning Commission's (CPC) revision of the cabaret law zoning regulations in 1990, which is discussed in the next chapter. This chapter examines the O'Brien test, which was the lynchpin of the *Chiasson* case, in which the NYC government failed to prove the legitimacy of the cabaret law regulation of live music venues according to the number of musicians and types of instruments in relation to its alleged aim of controlling noise and crowds. This chapter also reviews arguments made by Local 802 in the court about the historical legal disconnection between jazz music and social dancing, and how changes in social and judicial perception of particular kinds of entertainment, such as jazz music, often helped to rescue these kinds of entertainment from stringent regulatory governance. These points are revisited in Chapter 8 when I analyze the 2005 lawsuit that dancers filed in order to challenge the cabaret law in terms of its discrimination against social dancing (*Festa v. NYC*), as similarities in these two cases demonstrate important aspects of the licensing and zoning systems that govern entertainment in the cities, as well as the possibilities and limitations of grassroots movements against this system.

Third, this chapter also examines the ways in which the presence of artists and spaces signifying the subcultural richness of socially deprived neighborhoods, such as clubs, triggered the "hipster gentrification" of these neighborhoods. This period corresponds to that of "second-wave gentrification" (Hackworth and Smith 2001), in which well-educated upper-middle class professionals moved into the inner city interested in the artistic/cultural avant-garde and the mixed-use diversity that the inner-city offered. And, nightlife businesses, the history of gentrification reveals, were actually encouraged by the gentrification regime—by agents both in the real-estate market and also by governmental officials at both municipal and state levels—to raise the appeal of the area as pioneers of gentrification. However, once gentrification kicked in, nightlife businesses were blamed for defiling the quality of life of those neighborhoods. This shows the flaws in contentions by some urban scholars that gentrification (and by implication, the post-industrialization of urban space, economy and demographics in the city) actually revives radical urban creativities and leads to productive and progressive social/cultural mixing.

NEOLIBERALIZATION, POST-INDUSTRIALIZATION AND GENTRIFICATION IN THE 1980S

As was examined in the previous chapter, the direct or indirect takeover of the city's governance by financial elites had become institutionally entrenched in the process of coming out of the fiscal crisis period in the late 1970s. Under this institutional pressure and the disciplinary monitoring of the MAC (Municipal Assistance Corporation), the policy platform of the post-fiscal crisis mayor, Ed Koch, articulated that it would run the city primarily on a business basis, a priority which would, as we will see, be echoed by succeeding mayors (Brash 2011: 33). Budget balancing had settled as the most important policy objective of the administration, which, with the reduction in federal outlays to municipalities, spurred neoliberal roll-back policies consisting of cuts in public service and social welfare, while not increasing taxes (ibid.; cf. Peck and Tickell 2002). While lower echelons of society, especially racial and ethnic minority groups, were suffering from the loss of (manufacturing) jobs, job markets characterized by increasingly precarious employment and cuts in welfare benefits and public services, the priority of the municipal spending was placed in raising the appeal of the city in the context of intensifying inter-urban competition in order to attract (global) private capital, corporations and related workforces. Mayor Koch pioneered the corporate largesse system—bundles of benefits granted to corporations to lure or keep them in the city, a strategy which succeeding mayors also followed faithfully (Fainstein 1994).[1] The mayor also sought the post-industrialization of the city's economy, and the size of the professional-managerial class in the city reached new levels (Mollenkopf and Castells 1991). The city's dependence on two highly volatile, crisis-prone post-industrial businesses—that is, finance and real-estate—also deepened.

The NYC property market started to revive from the late 1970s, and a wide range of areas in Manhattan began to experience gentrification. Gentrification in the context of NYC in the late 1970s promised several advantages for the municipality. A workforce associated with post-industrial businesses that would move into the city through gentrification would not only revive the city's depressed economy, but also further facilitate the post-industrialization of the city's economy and space. As developments in the financial and the real-estate market became increasingly interwoven with each other, the deregulation of financial markets would open the door for foreign capital to invest in the domestic real-estate market (Hackworth and Smith 2001). Gentrification, as an investment opportunity, would also bring in (global) private capital into the city's real-estate market, which had been reluctant to enter the city. Gentrification would also bring significant revenue to the city's depleted coffers through the collection of property taxes. In addition to these economic gains, gentrification would also offer an impression that the municipality was putting in an effort to "sanitize" crime-ridden inner-city areas, and restore derelict built environments within them, through which the municipality could acquire legitimacy among the populace. These advantages of gentrification moved the city government to commit itself to establishing the political and economic conditions hospitable to gentrification.

Starting in the late 1970s until the end of the 1980s, capital investment (especially in housing stock) was concentrated in areas that Hackworth

(2001: 866) has called "a reinvested core (RC)," which refers approximately to the Manhattan area below 96th Street and the northwest part of Brooklyn (see Figure 4.1). With a few exceptions, most inner-city neighborhoods within the RC experienced severe divestment during the 1960s and 1970s. Starting in the late1970s, however, these abandoned neighborhoods became the "new frontiers" of gentrification (Smith 1996). Abandoned inner-city areas were very often targeted by the development sector and public institutions as sites for gentrification projects, with the "rent gap" (Smith 1996: 67) often cited as the economic force that initiated inner-city gentrification. As suburbanization started to show falling rates of profit by the 1970s, capital's attention turned to other avenues of investment (ibid.:

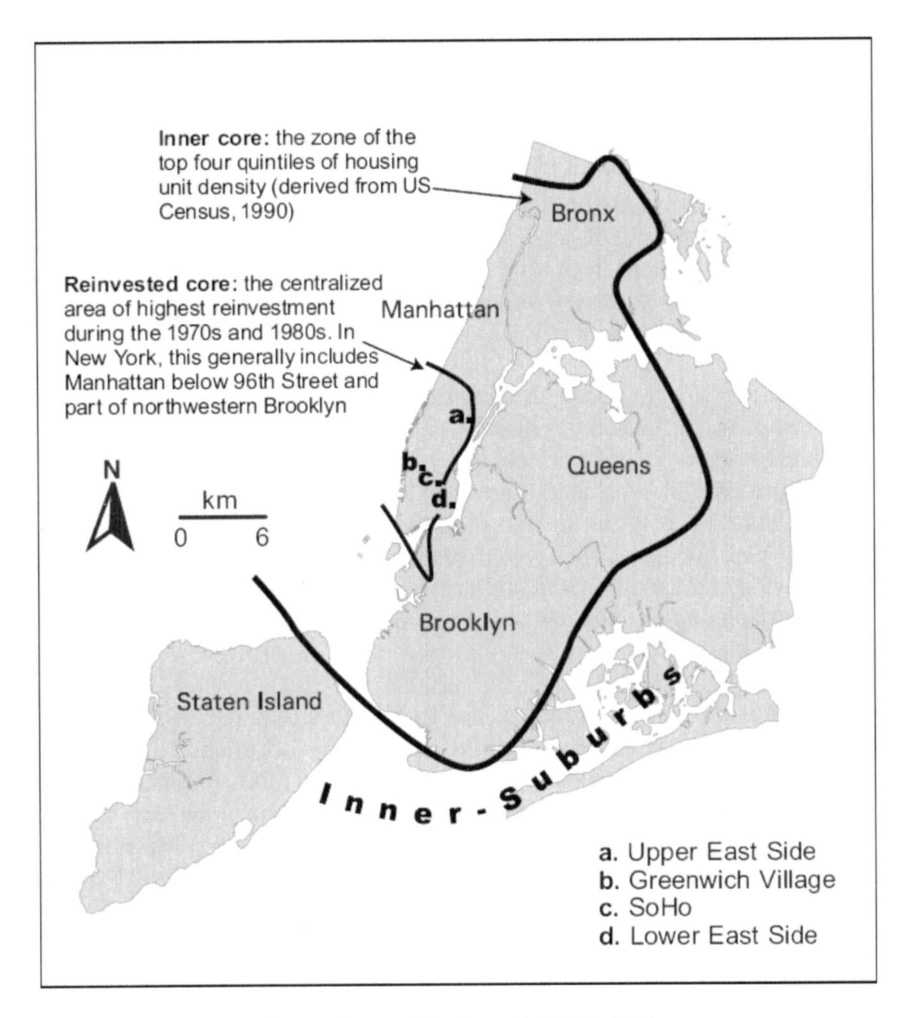

Figure 4.1 Reinvested core. Source: Hackworth (2001: 865).

75–91). Derelict inner-city areas looked promising as an alternative site for capital investment due to the deflated land prices there, and gentrification occurred when the rent gap was sufficiently wide for capitalists and developers to feel confident about purchasing structures cheaply, paying low amounts of interest on mortgage and construction loans, and selling the end product at a sale price that left them with a satisfactory return (ibid.: 68). However, investment did not automatically occur even though the rent gap grew sufficiently enough, as investors were still afraid of risky factors remaining in derelict areas. The presence of interesting subcultures in these derelict neighborhoods often became a factor that lowered the risks of entering into the real-estate market in these neighborhoods (Mele 2000). This is because investors often saw the presence of such subcultures as raising the appeal of the neighborhoods, as these subcultures were compatible with the cultural capital of potential upper-middle class gentries. These neighborhoods included SoHo, TriBeCa, the East Village (formerly called the Lower East Side), and the Flatiron District.

By the middle of the 1980s, the gentrification of inner-city areas in Manhattan had progressed a great deal. There was a significant drop in dilapidated vacant housing units in the city between 1981 and 1984, and during the same time period, rents for substandard housing rose a staggering 20 percent (Mele 2000: 223). Gentrification in the RC area did not just mean the resuscitation of abandoned residential properties, but entailed broader changes:

> By the 1980s, the social and functional heterogeneity of southern Manhattan was noticeably reduced . . . an uncounted number of factories had disappeared or had been converted to other uses, and large expanses of proletarian tenements had been replaced by expensive apartment towers. Chic restaurants occupied abandoned factory showrooms. The fabric of the central business district had changed: many strands of its previous industrial woof had been exchanged for the golden threads of late capitalism. (Fainstein and Fainstein 1989: 59; cited in Hackworth 2001: 866)

"The golden threads of late capitalism" added to the residential "upgrades" in formerly dilapidated areas included state-of-the-art office buildings that housed young professionals who were engaged in FIRE (financial, insurance, and real-estate) professions, or were employed in high-end services or culture/media industries. It also included the symbols of cultural revitalization of the central city, such as (high-end, or corporate) cultural amenities, shopping, entertainment and leisure facilities, and world-class fine art districts, such as SoHo. The tourism industry also enjoyed a great deal of success, with revenues up from $3.3 million in 1975 to $17.5 million in 1987 (Sleeper 1987: 438). Commercial space was aggressively redeveloped, with roughly forty-five million square feet of new commercial space built in Manhattan between 1981 and 1987 (ibid.), and BIDs were introduced to revitalize depressed commercial districts (McArdle and Erzen 2001).

Private-public partnerships were key in developments such as Battery Park City, The Metrotech project, Westway and the 42nd Street (Times Square) redevelopment project. These partnerships often took shape through the mediation of bodies such as the Public Development Corporation (PDC, which after 1992 was called Economic Development Corporation [EDC]), a nonprofit local development corporation which was operated in an entrepreneurial fashion with business-dominated boards (Fainstein and Stokes 1998: 164). It liaised between the municipality and the private sector in particular development deals, and designed incentive packages for private sector participants, such as bundled combinations of tax reductions, public subsidies, site improvements, zoning variances and the like. Multiple actors from civil society criticized these deals as boondoggles that were exhausting municipal coffers (Sleeper 1987: 439). As the PDC enjoyed considerable autonomy, and was not subject to the city's Uniform Land Use Review Procedure (ULURP), criticism was also raised in relation to the serious lack of public oversight and direct political accountability in the projects that PDC developed (Brash 2011: 40). Development projects under this body were promoted on a project-by-project basis, which steered the municipality away from the directive of long-term, comprehensive and coherent urban planning (Brash 2011: 40; Sleeper 1987: 437). The property bubble fueled by speculative and unplanned overbuilding eventually ended in a bust, two years after the stock market crash in 1987.

The gentrification of inner-city neighborhoods has brought about a severe class war, due to the class-biased nature of those projects. Most of the municipal inner-city revitalization projects were geared to assisting profit-driven housing schemes targeting high-income classes. At the same time, public subsidies in affordable/social housing for people who were increasingly dislocated to the economic margins with neoliberalization and post-industrialization of the city were cancelled. The HPD (the Department of Housing Preservation and Development), created in the early 1980s to protect low-income neighborhoods from the pillage of disinvestment, soon became an agent that assembled city-owned properties, which had often housed public facilities used for social programs, and provided them to private developers at attractive prices (Smith 1996: 70). The rampant upmarket conversion of Single Room Occupancy (SRO) hotel units further removed a form of housing from underclass citizens and homeless people which had previously been accessible to them (Sleeper 1987: 447). The municipality also provided incentives to landlords to renovate their abandoned housing (e.g. tax policies such as MCI, J-51, or 421-a), which resulted in rent hikes within these buildings due to the way these incentive programs were structured (for more details, see Mele 2000: 241), once again removing this housing from being affordable to low-income or people. The mayor's arguments, which he made often, that the private market would provide a better housing arrangement to low-income citizens proved to be blatantly misguiding (Sleeper 1987: 439).

The other pro-gentrification policy of the Koch administration was that of Quality of Life policing, which was implemented in order to clear out the "nefarious" street people and cultures that apparently stood in the way of area "revitalization." The Koch administration's clampdown on homeless people (in Tompkins Square Park), anti-loitering initiatives, a crackdown on street vendors and drug busting targeting minority youths, were some examples (Mele 2000: 239; Sleeper 1987; Smith 1996). This punitive form of policing was also extended to marginal, low-profile businesses that used to prevail in derelict commercial and manufacturing neighborhoods. An example of this was the peep show theaters of Times Square—once one of the defining types of commerce in the area—that became targets for closure, as the redevelopment of Times Square focused mainly on office complex developments. Gentrifying residential districts also witnessed neighborhood struggles over the presence of "unwanted" businesses. These businesses, due to their incompatibility with residential land-uses, were exerting numerous nuisance effects over neighboring residential areas, ultimately provoking a fear of depreciation of property values among gentries and local economic development associations. Cabarets and other nightlife businesses were one of them, and they became one of the main targets of Quality of Life policing.

CONFLICTS OVER DANCE CLUBS AND TIGHTENING OF CABARET LAW ENFORCEMENT

At the end of the 1970s, after-hours clubs as well as mega dance clubs continued to be financial boosters to the city, but as Horsley (1978) has reported, the "carnival atmosphere" that they created—noises, crowding, vandalism and behaviors that strongly indicated that clubs and their patrons were at the borderline of legality—increasingly provoked concerns from officials, urban planners, and residential community groups. For example, Community Board (CB) 6 persuaded the Board of Estimate to revoke an application for a zoning variance to convert a bar into a discotheque in the Murray Hill neighborhood on the grounds that this conversion would exert adverse impacts on surrounding residential neighbors (ibid.).[2] On the Upper East Side, where disco became "a dirty buzzword for communities" according to a member of CB8 (Horsley 1979), even a roller skating facility met with furious protests by residents, who argued that a roller rink would play disco music, and would attract typical disco crowds who would make noise and cause problems. In addition to residential communities, licensed clubs also complained about discotheques and after-hours clubs that frequently violated laws by avoiding tax duties or running as private clubs (by having a membership program) in order to avoid complying with required regulations, and enjoying an "anything goes" ethos, including open drug use. Despite these complaints about the operation of dance clubs, proposals for locating discotheques were relatively easily accepted by the Board of Estimate. The proposal's argument that the rejection of the proposal would deprive owners of their considerable investment and hurt the city's economy appealed to the Board (Horsley 1978).

Yet, the city soon started to respond to the disco dilemma in an increasingly uncompromising manner. "Disco owners have had a field day in licensing and there is a question of the weakness of government," said Councilmember Henry J. Stern (Horsley 1979), who had proposed an amendment to the cabaret law in January 1979. Following this initiative, the CPC, in collaboration with the Department of Environmental Protection (DEP), started to develop a new cabaret law regulation that would more effectively control the noise and crowds that were caused by discotheques. The cabaret law at that time did not have a specific zoning rule for discotheques. As explained earlier, discotheques belonged to Use Group 12A, which also included live music venues that allowed more than three musicians and/or music played by horns and percussion. Since discotheques had a much greater nuisance impact on surrounding neighborhoods than other nightlife businesses categorized under Use Group 12A, the CPC deemed it necessary to separate regulations for discotheques from those for these other businesses, and subject discotheques to much stricter zoning restrictions. The suggested amendment of the cabaret law would require special permits for all new dancing facilities in *all* planning zones where discos were allowed to locate. Under the special permit system, the application of the cabaret license by a disco owner would be reviewed by the Board of Standards and Appeals, which would consider whether the applying disco would be compatible with surrounding land uses. The special permit requirement would include a waiting-space requirement, "to cut down on boisterous crowds blocking city-owned sidewalks" as the CPC counsel said (Horsley 1979).[3]

In the midst of the emerging neighborhood disquietudes over discos, the municipality moved to secure safety inside cabarets. In 1979, a fire-safety law, called Local Law 41, took effect to cover the city's nightclubs. The law was initially prompted by a fatal fire that occurred in December 1975 at a nightclub called Blue Angel,[4] located in midtown Manhattan, but the law was not signed until 1979. This law required automatic sprinklers, fire alarms and emergency lighting in cabarets with a capacity of over seventy-five persons. Local Law 41 was one of the building-safety laws legislated between 1979 and 1983 (Dunlap 1983). The administration's move to legislate for building safety must have been motivated by the easing of the fiscal crisis and the fact that the administration now had financial resources to oversee the compliance of laws by the owners of buildings that had been derelict due to neglected management during the fiscal crisis. It could also be the case that as gentrification became the main goal of the city government, the city government may have felt it urgent to upgrade the built environment in the city.

There was another point of debate with respect to zoning regulations of discotheques, which was illustrated with the episode related to Private Cabarets Inc. which attempted to open "cabaret-type facilities" in the Yorkville neighborhood. Furious community opposition developed when the word spread that these facilities consisted of a Studio 54-type discotheque that would attract great numbers of patrons. Pointing to the problem that communities tended to typecast all dance venues as big mega discos, the attorney of Private Cabarets Inc. argued that the city's zoning resolution was also making the same mistake:

> A disco is not defined in the zoning resolution. The resolution fails to recognize any distinction between a Copacabana-type nightclub and a neighborhood bar where a happy patron can dance a jig to a record on the jukebox. (Horsley 1979)

This inability of lawmakers and communities to distinguish between these two forms of entertainment would have serious implications for the city's regulation of cabarets from 1990 onwards, a point that will be further discussed in the ensuing chapters.

In 1982, as the fiscal crisis eased further, the DCA was given money and power to shut down clubs that violated the law, without mediation by courts (Chevigny 1991: 81). The practice by which officials had turned a blind eye toward music-only clubs, or illegal businesses that were apparently harmless, also began to change by the early 1980s (ibid.). The DCA not only tightened nightlife inspection, but also increased fines for violations. The liability that owners of nightclubs had to incur in cases of building code violations also grew disproportionately. Owners who violated the law used to face a maximum of $500 in criminal fines and up to six months in prison. In 1982, the city added $5,000 to the maximum fine and an additional ninety days to the maximum sentence (Dunlap 1983). By 1983, it was reported in the *New York Times* (*NYT*; ibid.) that the city's cabarets felt that the permissive atmosphere that characterized the 1970s was gradually vanishing. City officials said that the number of nightclubs that complied with the law had indeed increased, due to stepped-up enforcement and increased penalties. The *NYT* article reported the protests registered by owners of small theaters, as the requirements of Local Law 41 had also been applied to small and makeshift theaters that often screened shows in abandoned buildings. A representative of a trade organization for Off Off Broadway theaters and theater groups asserted that the sprinkler requirement was "truly onerous." This requirement forced these Off Off Broadway theaters to opt for one of the following: (a) closing the theater; (b) installing sprinklers to buildings that theaters had no equity in (and landlords would not invest in this equipment); or

(c) reducing the capacity of theaters to seventy-four persons so in order to avoid being under the purview of Local Law 41. It can be inferred from this report that such fire safety requirements must also have been onerous to small music/dance clubs and parties.

The special projects director in the FDNY's Division of Fire Prevention stressed the urgency of monitoring the safety compliance of nightclubs that featured entertainment. "They are volatile environments," he averred:

> [T]he average person in a cabaret hasn't the vaguest idea of where the exits are . . . People are drinking, sometimes there are funny cigarettes going around, there are high noise levels and people are getting angry—attempting to adjudicate their grievances in a violent form. (Dunlap 1983)

This type of discourse was very much characteristic of how the authorities understood nightclubs, and such statements were quite often submitted to justify moves to tighten safety rules in nightlife businesses that offered entertainment. Fatal fires that have occurred in nightclubs—such as those in Blue Angel in 1975, and later on, in Happy Land in 1990, the latter of which I discuss in Chapter 6—also have been submitted as evidence that reinforced the authorities' conviction of the danger inherent to the nightclub environment. According to Chevigny (1991: 82), approaches taken by authorities of treating nightclubs as volatile environments have been successful in persuading courts to strike down challenges made by nightclubs against stricter safety rules. Even though both fires mentioned above were fatal because the two sites involved were not equipped with basic safety measures proper to their *sizes*, not because nightclubs were more volatile environments than any other nightlife businesses, this explanation never gained much circulation. The law was indiscriminate in applying the same standard to all sizes of businesses that had any magnitude of entertainment on offer, and therefore inimical to small businesses.

In April 1983, the City Council's Consumer Protection Committee proposed a bill that would require establishments generating loud noise to be soundproofed. The head of the Committee said that cabarets, discos and record stores should acquire soundproofing certificates in order to remain in business. She continued:

> I don't know of any elected official, particularly those who represent Manhattan residents, who does not constantly get calls from people who live near cabarets, discos and even certain restaurants complaining about the loud music that blares forth . . . This particular problem is exacerbated by the fact that these establishments tend to be open very late during the exact hours when people are trying to sleep. Some rock sounds have a penetrating beat that can be more annoying than the music itself. (Kennedy 1983)

She added that laws aimed at controlling noise were becoming increasingly important because the distinction between commercial and residential areas had been blurred, indicating that gentrification of commercial, or mixed-use districts lay at the heart of problem.[5]

With tightened regulations on cabarets being implemented throughout 1982 and 1983, small live jazz music clubs that intended to have more than three musicians and allowed horns and drums to be played in their venues suffered tremendously, as they all were included in the same regulatory category (Use Group 12) under the cabaret law as mega discotheques, and thus were subjected to the same degree of regulation. Owners said that the estimated costs of installing the required safety devices and abiding by the other stipulations would range from $25,000 to $50,000, on top of license fees of $800 and $1,000 depending on the venue's capacity (Freedman 1986; Pareles 1987). Left-leaning Councilmember Ruth Messinger crafted a bill in 1983 in collaboration with the Musicians' Union Local 802 to find a breakthrough in this impasse for small live jazz music clubs. This bill proposed to eliminate the discrimination of the cabaret law against horns and percussion, and also proposed to limit commercial rents for small businesses. This proposal (especially the second part of it) invited the enmity of the Koch administration, and the bill was never passed as law (Chevigny 1991: 87). The chairperson of the Committee on Consumer Affairs insisted that the Messinger bill, if passed, would allow 14,000 restaurants in the city to have jazz combos, and since many of them were in residential zones,[6] they would invite unceasing complaints from residents (Freedman 1986). For Local 802, this was not a plausible argument because by that time, any live music could make a roaring noise with electronic amplification technology, so it made no sense to regulate only jazz combos whereas the other music instruments, which may well produce more noise, were allowed in restaurants and/or bars in light commercial zones. In response to the Committee's reaction, in 1984 Local 802 started to draft another bill with some council members that would propose to limit the decibel level of music played inside clubs in light commercial zones, regardless of the instruments used.

In 1984, the city granted the DCA more funds to hire more inspectors and consequently the enforcement of the cabaret regulation became more thorough (Freedman 1986).[7] Live jazz music establishments started to respond by reshaping the feature of the music according to the changed conditions. Piano and bass duos became routine, due to the restrictions imposed on the number of musicians and types of instruments (Chevigny 1991: 82). Some jazz music club owners also responded to this situation by applying for a cabaret license, but many of them found this to be quite a hard process. Lush Life, located at the border of Greenwich Village and NoHo, applied for a cabaret license after having successfully created jazz performances, only to give up the attempt in 1984 after discovering that the application process would not be completed for at least three years. The club's owner complained:

> unfortunately, the bureaucracy of New York City does not see itself in a supportive role to help well-meaning people who want to comply, to make it possible to do it. The standards are contradictory. It's a miracle that anyone has been able to comply with the requirements in New York. It's *very* costly. (Chevigny 1991: 83, emphasis in original)

In 1985, the Consumer Affairs Committee in the City Council held a hearing on the Messinger bill together with the noise bill that Local 802 drew up. Local 802 insisted at the hearing that the noise level had nothing to do with the types of instruments played. Various residents and community members came to the hearing to complain about noise in their neighborhoods, but most of the complaints were more about raucous bars or discotheques rather than about live music premises (Chevigny 1991: 88). Eventually, the noise bill was passed, whereas the cabaret reform bill remained trapped within the committee, and had little possibility of being passed in the near future. Chevigny, who was present at the hearing, recollected:

> it [was] just very difficult to convince people that important cultural resources may be offered in bars and restaurants that are the entertainment equivalents of Mom-and-Pop stores." (ibid.: 89)[8]

Politicians and administrators were apparently indifferent about preserving small music businesses. Chevigny (ibid.: 89) also realized at that point that the cabaret reform would not be accomplished without zoning changes, because licensing of cabarets presupposed zoning regulations. The Messinger Bill, if passed, would bring only limited deregulation of the cabaret law because it did not incorporate a proposal for zoning reforms. However, there was no political road to amending the cabaret zoning system, and by this time, it had become next to impossible for groups like Local 802 to move the CPC to revise zoning resolutions, however modest the proposed revisions would have been. The propertied saw any modification of the zoning resolution of the cabaret law—or any modest changes in zoning system that involved "nuisance" land-uses—as a sign of trouble starting in their neighborhoods, and politicians feared provoking their major voting constituents. Chevigny realized at that point that what was needed was not better arguments, but a forceful countervailing political power to politicians and administrators, which Local 802 lacked at the time. This line of thinking led Chevigny to consider going to court.

THE PARTY WAS OVER

An *NYT* article in 1985 declared that "the party seems to be over for Lower Manhattan Clubs" in what is now called the East Village (Gross 1985).[9] Several reasons could be inferred for the blight in lower Manhattan's clubland. Some clubs were on a downward spiral due to newly emerging competitors; and the depressed club market in the area also suffered from the new law that altered the legal drinking age in New York from 19 to 21. In the meantime, rents continued to increase rapidly, and impose heavier burdens on small clubs—there was a reported increase in rents of 35 percent citywide over the period 1983–1985, and those whose leases had expired in that period reported an increase of 66 percent (Chevigny 1991: 98). In addition, the increasingly tightening administrative grip on clubs had a crippling effect on the area's clubland.[10] Darinka, 8 B.C. and the Limbo Lounge were hurt by the complicated procedure of cabaret law applications. King Tut's Wah-Wah Hut which staged performances once a week failed to acquire a cabaret license due to the cost of conforming to building codes and obtaining a zoning variance. The Peppermint Lounge, famous for popularizing the Twist, was battered by the government's crackdown on organized crime.[11]

8 B.C. (located on 8th Street between Avenues B and C), was a typical smoky late-night dance hall and demimonde cabaret that featured art and performance together with social dancing—a place that reflected, according to one of the owners of the club, "the disordered energies that were being flung with wild abandon on [the East Village] in the last two years" (cited in Gross 1985). Soon after the club opened in 1983, it became entangled in a stream of violations. The owners insisted that they had tried to comply with the numerous laws involved in the opening of a nightclub: "License fees, certificate of incorporation, lease or bill of sale, nonflammability of drapes, floor plan and building approval, public assembly permit, fire inspection, electric and gas inspection, health permit, liquor license, consumer affairs" (ibid.). However, since the club was located in either an R7 or a C1–5 zone (judging from Figure 2.1), meaning that it was off-limits to cabarets, the club had to apply for a Certificate of Occupancy to convert the current use to Use Group 12, as well as for zoning variances. The club owners eventually failed to acquire permission, even though the club met nearly all other requirements of safety and occupancy necessary for a club's operation.[12] In September 1985, the club was cited as an illegal cabaret, and five days later, the two owners signed an order that finalized the closure of the club.

The story of 8 B.C.—like many others related to the city's demimonde cultural businesses—presented an interesting irony in relation to urban revitalization. The municipality promoted the Lower East Side Cross-Subsidy Program in July 1984 to revitalize the abandoned Alphabet City (the areas of Avenue A, B and C) by selling vacant city-owned buildings. An assistant commissioner of the Department of Housing Preservation and Development (HPD)—the municipal institution at the vanguard of gentrification at that time—said that clubs like 8 B.C. were looked upon positively by HPD because they were considered to be resources that would bring people into derelict neighborhoods, eventually making the area an economically viable community (ibid.). 8 B.C. was one of the pioneers of gentrification of the neighborhood, as the assistant commissioner seemed to be aware of. While the importance of clubs in regenerating communities was acknowledged, it was deemed out of the question for the municipality to offer serious thought about these clubs as important cultural institutions and anchors of certain subcultural communities.

Instead, the authority's reaction tended toward regulating those clubs *de jure*, and sweeping them away as petty nuisances that defiled the neighborhood's up-and-coming character. The authorities' crackdown had particularly detrimental effects on small, more creative venues. Limbo Lounge, a combination of art gallery and nightclub, was located in either an R7 or C1 zone, where cabarets were forbidden. The owners tried to go legal (they contemplated getting a zoning variance, like 8 B.C. had), but it did not take long for the owners to learn that the chances of gaining a cabaret license were slim at best. The owners got rid of the nightclub business and changed the premises to Limbo Gallery and Theater. Another example was Darinka, a performance studio that held no more than 100 patrons, and where the owners started to sell liquor to patrons because they found it impossible to keep up the business without doing so due to the rising rent. The club was busted as an unlicensed "bottle club," and closed down in July 1985 after being in business for a year and a half.[13]

These small clubs were also occasionally terminated by landlords at the end of their lease, in cases where landlords did not want these clubs to continue in their buildings. These clubs, with the cooperation of landlords who, when it was convenient to them, purposefully overlooked their illegal operations, often operated in old, decaying buildings that did not abide by the building code. Vito Bruno, the owner of Pizza A Go-Go and also a persona that promoted outlaw outdoor parties for the first time in New York said, "The kids can't afford to get the buildings up to code, which is the landlord's responsibility anyway. But then the landlords want them out because they want to upgrade" (ibid.), even though these areas had often become revitalized only after nightclubs and other subcultural businesses had attracted pedestrians and helped breath life onto once desolate streets.

Live jazz music clubs were heading towards a similar destiny. By the end of the fiscal year 1986, Local 802 had identified twenty-five clubs that were eventually closed down by the authorities (Chevigny 1991: 18). Most of these clubs were obscure and small, but somewhat better known (but still economically marginal) clubs were also affected. West End Bar on Upper Broadway on 113th Street, Angry Squire in the Flatiron neighborhood and others were fined or padlocked (ibid.: 83). For many small jazz clubs, getting cabaret licenses was very difficult, for reasons that dance/music/performance clubs were also struggling with. The owner of Angry Squire said:

> I have commercial zoning, and I'll pay my annual whatever for a . . . permit. But it's hard to put thirty grand into this small place. Why should I have to have the same license as a big dance hall with

> thousands of teenagers? This is a listening-only club with seventy seats . . . (Pareles 1986)

Bracketing small live music clubs and large discos in the same regulatory class caused serious cost problems for small clubs, but the problems did not cease there. For many clubs, it seemed almost impossible to figure out how long it would take to receive the license after they submitted their application, and whether the license was actually going to be granted. Some clubs eventually decided not to have live music at all because having it would entwine them in expensive and cumbersome bureaucratic hassles. The hardship experienced by the clubs trickled down to musicians, too, in terms of their employment and pay. The economic hardship and the disappearance of cabarets meant that there would be fewer establishments where musicians, especially those who played horns and percussions, could make a living (Chevigny 1991: 84). Clubs were also in a catch-22 situation, as getting rid of music meant the loss of patrons, while keeping or adding music carried the high likelihood that it would result in additional costs due to bureaucratic processes (ibid.: 147–48).

THE MUSICIANS' UNION'S LITIGATION

The Chiasson Litigation

Eventually, Local 802, in collaboration with Paul Chevigny, decided to go to court in order to challenge the constitutionality of the provisions of the cabaret law. The process of this litigation is detailed in Chevigny's book, *Gigs: Jazz and the Cabaret Laws in New York City* (1991), which I draw upon in the account that follows, here. Chevigny reflects that the chances of winning the lawsuit at the time were slim because legal awareness of live music as a right protected under the First Amendment as "freedom of expression," was only just beginning to surface both inside and outside of courtrooms; until the early 1980s, neither jurists nor musicians themselves considered the rights of live music and musicians in terms of constitutional rights (Chevigny 1991: 6). In addition, in lawsuits concerned with the city's cabaret law up to that point, judges had never struck down the cabaret ordinance according to constitutional principles, even though they often questioned the constitutionality of the law as it was enforced upon live music premises (see examples of these lawsuits in ibid.: 114). Given that the question of the constitutionality of cabaret laws seemed to be up in the air, it was likely that the courts' decisions would be affected by the degree to which the judge cared about the importance of live music played in nightlife businesses.

Despite the odds against them, the musicians' frustration with the cabaret regulatory system was such that the voices for change through litigation increasingly gained ascendancy among them. Framing live music as a constitutional "right" supplied a robust and empowering framework through which musicians could articulate the problems of the cabaret law and the importance of their work (ibid.: 6). This is an example that shows how instrumental rights talk is in spawning (new) consciousness of particular entitlements that have not yet been conceived of as such, as I discussed in the Chapter 1 of this book. The framework of rights, as is shown in this case, helps victims to define what is undue and unfair, develop a new collective identity and solidarity among themselves, and make a coherent claim about their entitlements. With the key argument that live music is a right, and that the cabaret law represented a violation of free expression, Local 802 finally filed a lawsuit in May 1986, with a motion for an expedited hearing through a preliminary injunction. In what became *Chiasson v. New York City Department of Consumer Affairs* (1988, from now on *Chiasson*), Local 802, represented by Paul Chevigny, argued to the State Supreme Court that the cabaret law, as it was applied to live music played in eating and/or drinking establishments in the city, was unconstitutional as it violated the freedom of expression of musicians.

The city's rationale for the cabaret regulations of live music venues was that the presence of more musicians and of live music performance including horns and percussion invited more noise and traffic congestion into neighborhoods; therefore, they should not be allowed in light commercial zones where residential uses heavily existed and only Use Group 6 commerce—local service establishments to fill local residential needs, and ones that only provide "incidental" music—could exist. Local 802 contended that this rationale was not valid due to technological advances that could easily amplify any live music. Having deliberated for approximately one month, in June 1986 the judge filed his preliminary opinion enjoining the discrimination against horns and percussion, acknowledging that live music is a constitutionally protected form of expression, and that the municipality's regulation based on types of instruments was unconstitutional. However, he accepted the city's regulations regarding the number of musicians (Chevigny 1991: 115–116). The city's crackdown thus continued in targeting clubs that allowed more than three musicians, effecting damages to clubs, as had been done before. Mikell's, for instance, was going into debt, and the owner told Chevigny that if the club could only offer trios, the owner would not be able to pull the business out (ibid.: 126).[14] In the context of such troubles, Local 802 decided to push the litigation further, in order to persuade the judge about the unconstitutionality of the three-musician restriction of the cabaret law.

The *Chiasson* plaintiffs worked to reinforce their argument that the cabaret law could not pass the O'Brien test. Under the O'Brien test, a law that restricts freedom of expression for the purpose of public health is still constitutional if it proves that it (1) is content-neutral, (2) leaves open ample alternative channels of communication, (3) serves a legitimate governmental objective, and (4) is narrowly tailored to serve the governmental objective. First, Local 802 alleged that the cabaret law's regulation of the number of musicians could not pass the content-neutrality test. As trios usually excluded horns or drums, the cabaret law restriction of the number of musicians was a *de facto* discrimination against black and Latin music, and was a violation of freedom of expression, as it circumscribed the instrumentation and the style of music through which musicians express themselves (ibid.: 112). In addition, the restrictions in the number of musicians reduced adequate channels of communication by depriving young, obscure musicians of spaces in which they could express their musical ideas before an audience, and leaving few alternative channels of communication. Licensed (mostly mid- to large-sized) places were more likely to book prominent musicians, rock bands, or disco parties out of profit calculations, leaving obscure and younger musicians with fewer opportunities to play in front of an audience. The only spaces where these musicians were given a chance to play were, then, small ones that were mostly located in light commercial zones, which were off-limits to cabarets. In the cabaret law's optic, there was no such thing as a neighborhood jazz club, in which gigs composed of obscure musicians play horns, drums and other instruments for casual diners or drinkers; it only envisioned a big-time nightclub that involved horns and drums that attracts large numbers of people from all over (ibid.: 97). Local 802 also reminded the judge that the development of jazz music could only be secured through live performances by musicians, so depriving musicians of the space for performances, particularly in NYC—the capital of jazz music—meant the sterilization of the genre itself (ibid.: 121).

In terms of the cabaret law serving a legitimate governmental objective, Local 802 focused on challenging the city's assumption that more musicians would cause more traffic. A noted engineer who used to be the director of Planning and Implementation for the city's DEP, opined in his affidavit in favor of plaintiffs that there was no relation between the number of musicians who play in a venue, and the degree of traffic in the neighborhood; for him, the latter was more dependent upon the size of the establishment (ibid.: 125). The plaintiffs further argued that the cabaret law was not narrowly tailored to reach its ends; if the traffic was the problem, then it would be more effective if the government used traffic regulation to control it, or limited the size of restaurants and bars in the neighborhood whose characteristics were closer to residential. Nevertheless, the municipality had pushed the regulation based on the nature of entertainment itself under the rubric of traffic control, which, Local 802 maintained, only made observers suspect that there was a disingenuous unspoken motive to the regulation—i.e. the long-rooted racist fear about "jazz crowd," and the related fear of dancing (ibid.: 96).

Then, the plaintiffs directed the judge's attention to changes in the history of jazz. They argued that live jazz music is played for listeners, and no longer for dancers, but that the cabaret law's provision still bracketed jazz music and dancing together in its zoning and licensing schema (ibid.: 69, 77). This was also not fair as dance clubs could easily afford the high expenses associated with the licensing requirements for cabarets that small jazz clubs could not (ibid.: 77). Another line of reasoning by the plaintiffs was that patrons attracted to jazz clubs were much more genteel than those attracted to dance clubs, which was echoed by the local media. Local press was incredibly supportive of Local 802's legal challenge of the cabaret law (e.g. Editorial 1986; Levy 1986; Pareles 1987). By then, jazz had become an accepted and elite form of music, being prominently featured in art commentary sections of major newspapers, in contrast to the besmirched reputation of jazz in the language of city regulations. Chevigny (ibid.: 115) argued that such media support contributed to building a powerful force which countervailed those of city administrators and city planners.

Chevigny (ibid.: 113) recollects that dance clubs at that time could not form a collective force, due to the cacophony and disagreements among owners moored in the competitive nature of the industry. There was not yet a coherent organization of dance clubs to which Local 802 could avail itself so that the dance club community could collaborate with the musicians fighting in the *Chiasson* litigation.[15] Even though Chevigny says that the *Chiasson* plaintiffs would have welcomed dance clubs owners' participation in its litigation, in hindsight their absence in the litigation offered an advantage, as Local 802 could argue for the historical separation of jazz music and social dancing. More significantly, putting forward music only to the court turned out to be fortunate, because in 1989 the US Supreme Court decided in *City of Dallas v. Stanglin* (1989, from now on *Stanglin*) that there is no substantial First Amendment (expressive) interest in social dancing. This decision meant that, if music and dancing had collaborated in the *Chiasson* case, there may have been an increased likelihood that the plaintiffs would have lost the case. Chevigny commented:

> [The *Stanglin* decision] is not surprising, if regrettable, since the time for the expressive value of social dancing had not yet come within the limits of the official perspectives, even if the time for vernacular music [jazz music] might have arrived. (1991: 113)

The legal system had to wait for the idea of social dancing as expression to take root. The *Stanglin* ruling has become a significant precedent for litigations involving social dancing, and it was also a point of legal reasoning in *Festa* in 2005. I will re-visit *Stanglin* in Chapter 8, when I analyze *Festa*.

The judge finally ruled in 1988 that the cabaret law's regulation of bars and restaurants on the basis of the number of musicians was unconstitutional. This created pressure on the municipality to rewrite the cabaret law, and the anticipation grew about to what degree the municipality would relax regulations over live music, and how the relaxation would change the city's nightlife.

Law, Space and Subcultures

Before wrapping up the analysis of *Chiasson*, I want to briefly broach a question that Chevigny raised in relation to the contradictory nature of law in bringing about social change. *Chiasson* and the subsequent revision of the cabaret law (which the next chapter investigates) granted more liberties to musicians about what they can play and with whom, and legally enabled jazz combos to be played in many more bars, restaurants, and clubs than before. Ironically, such changes in the cabaret law did not much improve the livelihoods of live music clubs. As a matter of fact, the economic circumstances of these clubs plummeted due to rapidly rising real-estate prices, a thorny point that the reform of the cabaret law did little to solve. Therefore, musicians may have been *legally* allowed to play in more places, but due to the much wider *political economic* forces of real-estate that actually shrank spaces available for live music, the ideal of the *Chiasson* litigation was not achievable despite the legal victory. This limited effect of the *Chiasson* litigation also points to the limitation that "rights-based" movements and litigations can have in bringing about social change in capitalist systems (cf. Scheingold 1974). I will revisit this question again in Chapter 7, in relation to two pro-nightlife formations that emerged in late 1990s.

SYMBOLIC INCLUSION OF NIGHTLIFE INTO REAL-ESTATE MARKETING

Over the course of the 1980s, lousy dance clubs, and boisterous nightlife establishments in general, were pitted against the interests of "desirable" communities and property values of neighborhoods. However, a closer look at the process of urban revitalization, as it happened in NYC, reveals a number of cases in which a thriving nightlife has ushered in and even constituted an essential part of the revitalization of neighborhoods. Nightlife businesses created the vibe of lively urban socialization in the neighborhood that attracted people from outside these derelict areas. In particular, those nightlife establishments in derelict neighborhoods related to inner-city subcultures made these neighborhoods more attractive to outsiders by creating a particular urban aesthetic and subcultural distinctions, helped to revalorize depressed property and trigger gentrification, and enabled landlords and real-estate investors to reap "monopoly rent" (Harvey 2003). Realtors and developers used such an aesthetic as a selling point for neighborhoods to the culturally savvy and cosmopolitan upper-middle classes, as well as to tourists. Therefore, culture partially explains why certain neighborhoods—among the many that experience widened "rent gaps"—become the focus of gentrification during particular times.

One *NYT* article (Hawkins 1988) reported that this was the case in the Flatiron District in Manhattan. The Flatiron District used to be a manufacturing district (M1–5 and M1–6) combined with heavy commercial businesses (C6) until the 1960s. With deindustrialization and the subsequent fiscal crisis of the mid-1970s, investments in properties in this area dwindled, and abandoned lofts were occupied by poor photographers who came to the area in search of spacious studios for work and living. In the early 1980s, when gentrification expanded from SoHo and NoHo, it reached this neighborhood, too, and in 1981, the zoning code of the neighborhood was changed to M1–5M and M1–6M to legalize the entry of more residential units. Those who had stake in the area's real-estate market even started to brand the neighborhood under the name of "Sofi" (South of Flatiron), to give a SoHo-like air to the area.[16] The neighborhood had several merits that explained gentrification there, such as its central location in Manhattan, convenience of public transportation that reached virtually any place in Manhattan, surrounding residential neighborhoods such as Gramercy Park and Chelsea and the availability of spacious loft apartments at reasonable prices (that is, more affordable than SoHo loft prices). The realtors interviewed in the *NYT* article, however, opined in a single tone that the most important factor for gentrification in the district had been the nightclubs and restaurants that had moved into the area a few years earlier in search of lower rents. These realtors commented that nightlife businesses had brought in "excitement and glamour" to the neighborhood, and "attracted a lot of attention to the area." Drawing young crowds at night, one realtor continued, granted the neighborhood "the prestige that usually presages gentrification." The area was, however, increasingly experiencing severe conflicts over the nightlife that was proliferating there, as I detail in the next chapter.

TriBeCa was a light manufacturing and wholesale trade district until the 1960s. But the character of the neighborhood changed as artists who could not afford neighboring SoHo's rents any more moved down to it in the early 1970s and created loft residences and performance spaces here, including dance clubs and parties (Noble 1989). On June 11, 1976, the city designated the neighborhood the Lower Manhattan Mixed Use District (LMM) to legalize illegal loft-dwellers (mostly artists), and also to protect existing retail, commercial and light manufacturing businesses. Following in SoHo's footsteps (i.e. artists conversing of lofts into work places or performance places, triggering real-estate investments in upscale loft development), real-estate values soared in TriBeCa, too, the neighborhood continually attracting professionals in their late 20s and 30s, while galleries and restaurants proliferated.[17] The booming real-estate market, however, came into conflict with the neighborhood's nightlife establishments. In 1983, it was reported that some residents were attributing the increased crime rate in the area (despite the citywide drop in crime rate at the time) to discotheques in the neighborhood (Landis 1983), including the Mudd Club, which was counted as a venerable cultural institution by downtowners. Residents argued that the influx of hundreds of drinkers and dancers that had increased with gentrification of the neighborhood was causing an unprecedented number of crimes. While nightlife establishments pioneer gentrification, gentrification, in turn, ushers in more nightlife establishments as landlords encourage nightlife businesses to locate in a gentrifying neighborhood, and while, also, more nightlife establishments tend to locate in burgeoning edgy neighborhoods to take advantage of newly present demographics of young gentries (Ocejo 2009). The history of the transformation of TriBeCa is reflective of this contradictory dynamic that nightlife has played in the gentrification of NYC neighborhoods.

The history of the transformation of the East Village in the 1980s took a similar course, but it additionally bespeaks to how artists and the arts are often complicit in the dynamic. According to Mele (2000), the gentrification of the East Village explicitly hinged on the aesthetics and ambience that the neighborhood's counter-cultural and bohemian artists had created in the neighborhood landscapes during the 1960s and 1970s, in the middle of the dereliction of this particular neighborhood (also see Chapter 3 of this book). Clubs that were the locus of these artists' prolific production and experimentation with various art styles, music, dance and performance became, in the 1980s, a place where artists marketed themselves and their work to the mainstream art market (Mele 2000: 217). In marketing their work, artists took advantage of the counter-cultural ambience that defined the East Village subcultural scene, eventually inventing a so-called "East Village genre" that had the aura of specific attitudes and affects associated with the East Village (ibid.: 226). The media's attention to this urban genre gradually changed the popular picture of the neighborhood from low and marginal to central and interesting. This repackaging of the East Village elicited the interest of the real-estate development sector in the neighborhood, which had been until then afraid of entering into the development of the neighborhood owing to the long-standing negative images associated with it. The aesthetics of the East Village genre was appealing to middle-upper class professionals, and the real-estate sector marketed it as an object of chic lifestyle consumption available in such trendy urban neighborhoods. The history shows how artists often play the role of "critical infrastructure" (Zukin 1991: 214) or "cultural intermediary" (Bourdieu 1984) that channels the artistic and cultural avant-garde into an important part of the post-Fordist lifestyle consumption.

The appropriation of inner-city subcultures by the real-estate market proceeded, ironically, as the actual counter-cultural spaces were being displaced from neighborhoods, due to rent inflation and Quality of Life policing by the municipality, which heavily regulated the "nuisance" effects of nightlife businesses that originally nurtured these counter-cultures. The contradictory approach to the problem of 8 B.C., an East Village club, by the Department of Housing Preservation and Development (HPD), as illustrated previously in this chapter, attests to this phenomenon. The indifference of the authorities toward subcultural sites was contrasted with the municipality's generous sheltering of newly emerging East Village artists through governmental assistance programs. In 1981–1982, the municipality proposed the Artists Homeownership Program (AHOP) to convert in rem properties into artists' housing, as the East Village was establishing its name as an alternative art site. This project was thwarted by furious protests from community activists who were suspicious that the municipality was acting on an unspoken motive to use the program to induce upscale housing development (Mele 2000: 238–39). SoHo's successful sellout to the real-estate market in the 1970s brought recognition among urban planners and public officials that obscure artists' workshop studios in abandoned quarters often helped to make the neighborhood appealing in the real-estate market.

Mele (2000) names this process "symbolic inclusion" and "material exclusion" of vernacular cultures in the sense that the East Village countercultures that once defined the place now only remained as a feel and an image that was projected onto the new East Village landscape, upon upscale bars, restaurants, boutique shops and galleries, while the actual subcultural communities, their creative energies and their spaces had disappeared. Such a contradiction, embodied in the displacement or transformation of subcultural spaces and nightlife under gentrification, raises a vexing question for studies like Florida's (2004) that stress the importance of culture-oriented neighborhood revitalization. As I have already discussed, these studies emphasize the importance of "authentic" subcultures, including authentic nightlife, in facilitating economic revitalization of neighborhoods and the post-industrialization of cities, but have rarely pointed out, let alone discussed, how gentrification and post-industrialization have, *pace* this vision, displaced these "authentic" subcultures and nightlife, a process that has popped up in many cities, including neighborhoods in NYC.

CONCLUSION

Nightclubs that had existed in formerly abandoned neighborhoods increasingly became a nightmare for new residents in these gentrifying neighborhoods. Rallying cries by residential communities against nightclubs intensified, and the municipality responded to this by stepping up its regulatory grip on the city's nightlife. While the thriving club scene and vibrant nightlife in general were often represented as interesting neighborhood landmarks in the real-estate market, when it came to the conflicts over the nuisances that nightclubs were causing, the only consideration at the table was about how to discipline nightclubs as good neighbors, and where else to move them when they did not conform to the changing character of neighborhoods. This process parallels municipalities' sponsoring of artists with workshops/housing subsidies as an anchor for future real-estate capital investment in dilapidated neighborhoods, only to later remove the subsidies to relocate artists elsewhere once gentrification kicks in. The difference is that while municipal sponsoring for arts has been very open and direct, such encouragement for nightlife has been subtle and indirect.

Here, I want to bring to readers' attention the complaints made by the representative of Private Cabarets Inc. in this chapter about the regulatory inability, *or* refusal of distinguishing between Studio 54 type mega discos and small, ordinary bars where patrons dance to music from jukeboxes. This inability would still inform the design of new sets of zoning regulations over dance venues in the revision of the cabaret law in 1990 that I discuss in the next chapter. That is, in this revision, large and small dance establishments would be bracketed together as one category, and subjected to stricter zoning restrictions, whereas different sizes of live music venues would be regulated differently. In this revision, moreover, the municipality would drastically reduce the locations where dance venues would be permitted without abiding by strict special permits requirements. With the city's small, under-financed nightclubs already in a precarious standing due to rapidly rising rents and increasing anti-nightlife campaigns, the 1990 rezoning of the cabaret law would add another critical blow to clubland. I turn to this revision in the next chapter.

5 Zoning out Social Dancing
The Late 1980s

By the late 1980s, the city's clubland had altered considerably. The municipality continued to trumpet the city's celebrated nighttime entertainment as an important selling point of the city. Nightclubs that were the engines of cultural innovation and creativity, however, were gradually disappearing or decamping to other boroughs, while glitzy versions had increasingly settled in as the dominant players in Manhattan. Nightlife was under disciplinary offensive from the market, the state and residential communities. In this chapter, I examine how these offensives and other factors, such as AIDS and changes in demographics in Manhattan that took place with the increasing post-industrialization of the city had gradually changed the nature of nightlife landscapes and cultures, ushering in the "gentrification of nightlife." I illustrate this process with the example of Paradise Garage, a legendary underground club that closed in 1987, the same year that a club critic declared the "death of downtown."

Another subject of examination of this chapter is a municipal move to revise the cabaret law zoning rules, in 1990—the culmination of the decade-long conflict over the city's cabarets and the "gentrification with and against nightlife," that was described in the previous chapter. In "abstract space," to use Lefebvre's (1991) phrase—the space emblematic of modern capitalism in which abstract exchange value is predominant—two interlocking processes have been key: "commodification of and through space," and "bureaucratization of and through space" (Gregory 1994: 40). Space should be divided, categorized, bureaucratized and thus subject to controls for effective governing, especially in accordance with on-going processes of capital accumulation. Once bureaucratized and under control, space functions as one of the conditions through which the broader economic, political, social and cultural life is regulated and made compatible with the current regime of accumulation. The commodification of space in neoliberalizing and post-industrializing NYC, exemplified by upscale and corporate residential/commercial developments, and the commodification and marketing of the city's vernacular cultural spaces, including countercultural spaces, as an important appeal of the city, generated contradictions in numerous forms, some of which erupted in the form of neighborhood conflicts over nightlife. The 1990 rezoning of the cabaret law can be understood as an institutional fix to this contradiction. That is, by reshuffling the spatial configurations of nightlife in the city, the continued expansion of abstract gentrified space could be secured.

Several points will be underscored in my examination of the 1990 rezoning. First, dance clubs had been seriously stigmatized as a serious threat to surrounding residential neighbors, and in the final draft of the rezoning of the cabaret law in 1990, the places that dance venues of *any size* could be located as-of-right were drastically reduced. Exploring new locations had always been important for clubs because the feel of the areas where clubs were located often defined the ambience of clubs and clubbing experiences. Due to the 1990 rezoning of the cabaret law, however, dance clubs were now seriously limited in terms of being able to venture into new, edgy places, and were instead forced to recycle existing spots. Second, I examine how the particular *unilateral* perception about dance clubs—considering most dance venues to be large "louts"—led to the significant exclusion (although this did not happen without struggle) of voices speaking out in favor of the value of spaces for social dancing, and also an *absence* of voices calling for the governing body's attention to protections of small dance venues as opposed to bigger ones. Third, I also point out how an expressive subculture's standing in the judicial interpretation of the First Amendment entitlement shaped the forms and parameters of zoning rules that govern it. That is, while live music, recognized as a constitutionally protected form of expression in *Chiasson* as discussed in the previous chapter, was regulated according to the size of the venues in which the music was being played, social dancing, which was not recognized as being protected by the First Amendment, remained subject to sweeping zoning regulations.

During this process, the municipality also showed resistance to deregulating live music venues despite the court's decision that instructed the municipality to do so. The municipality is routinely disinclined to make exceptions involving regulations in land-uses related to entertainment, as it can easily provoke the propertied class. This speaks to the particular nature of the zoning system that governs entertainment businesses in the city, and the limited potential of changes in land-use that can be brought by actors involved in entertainment and expressive cultures as well as other politically and socially marginalized groups. The zoning system is a code for, and enabler of, the power structure of a society.

THE DEATH OF DOWNTOWN, 1987

While the *Chiasson* litigation continued, other struggles concerning the rest of New York's nightlife carried on as well. In 1987, the *Village Voice*'s club critic Michael Musto wrote that New York was experiencing "The Death of Downtown" (Musto 1987). He recollected:

> Two years ago, even while the cynics smacked their lips and claimed it wasn't as good as Max's or the Mudd, when every night was so uncontrollably exciting . . . hell, there *was* a scene. The new generation was flagging cabs back and forth between three hot clubs, where all their zany creativity combusted into a three-ring circus of intoxicating fun and fashion. A random stroll into any of these places hit one in the face with a giant lemon meringue pie of glitter, innovation, and raw talent . . . (ibid., emphasis in original)

Creative energies and an edge of adventure bustling in clubs like Area and Danceteria, both defunct by this time, were absent from new clubs like Palladium, "where yuppies [threw] business cards at each other" (ibid.). To bohemian downtowners, Palladium, the 3,500-person capacity, glitzy, commercial, and new Union Square club, owned by the co-owners of Studio 54, was the symbol of the uptownization of the downtown scene (also see Gross 1985). The biggest force that lay behind this uptownization, Musto continued, was the high cost of living downtown, which had shut down bohemian joints in the area. Money was now the driving force in the downtown scene, in which creativity had traditionally been the priority. Landlords were keen to rent their premises to expensive stores and bars that could service Wall Street workers. And, there was a shrinking interest among "upwardly mobile urban dwellers" in the kind of clubbing involving "smoky late-night dance halls and demimonde cabarets where dancing and drinking, art and performance, video and film all mix"—the kind of clubbing that was characteristic of the bohemian downtown (especially the East Village) in the 1970s (Gross 1985). Musicians, performers and promoters who were interested in entry into the mainstream market took what used to be authentic and raw downtown excitement, and pushed it into chic galleries, MTV shows and on to the shiny dance floors of the latest downtown clubs, to entertain newly affluent young Americans who had recently become the dominant constituents of nightlife. What these affluent denizens were offered, however, were sterilized images of bohemia, upscale/celebrity-oriented versions of liminality and transgressiveness or a bohemian-themed ambience often enabled by mere shallow décor. The true bohemians disappeared, signifying the beginning of the "post-downtown" era (Musto 1987).

The year 1987 also witnessed the lamentable closure of legendary underground dance club, Paradise Garage (from now on, Garage), after a glorious ten-year operation in an actual former garage in the SoHo area (see Figure 5.1). To many DJs, dancers, and critics involved in the underground dance scene, Garage had been highly influential. While Studio 54 is the archetype of disco in popular memory, for those who took disco music, dance, and subculture more seriously, Garage represented the paradigm of the disco era.[1] A close offspring of David Mancuso's Loft, Garage was the contemporary inspiration for many of the (underground, mostly) dance clubs and parties that formed part of the 1980s downtown scene. Following in the footsteps of Garage, these underground clubs and parties developed their own dance subcultures, focusing more on the organic quality of music, dance, and performance, and less on the flamboyant display and conspicuous consumption that characterized uptown clubs (Berry 1992: 4–5). This underground downtown scene continued to operate with verve even after the mainstream disco fad declined in late 1979.[2] The main music genres that these clubs and parties were engaged in were house or garage,[3] but in the early 1980s, there were numerous efforts among DJs to accomplish crossovers between dance music and hip hop, rock, or reggae (ibid.: 57).

Like the underground disco culture of the 1970s, the music in these clubs was developed through subcultural institutions that were stationed in the downtown area, such as specialty record stores and independent recording studios. These key institutions were embedded within one another, and provided a crucial resource to the development of underground music and clubs. Patrons of these clubs were always mixed, but a significant portion of the contingent was urban, young, gay, black, and Latino men (Buckland 2002; Fikentscher 2000). As in the case of disco, this contingent formed not only the majority of the patrons, but also represented those who derived the most meaning and pleasure from participating in this type of clubbing. These

Figure 5.1 The former Paradise Garage at 84 King Street in SoHo in October 2006. Photo courtesy of Benedict Wallis.

clubs provided a kind of "cultural security zone," in which marginalized populations could enjoy dancing both as fun and as statements of identity, in which they could find sanctuary away from the mainstream, homophobic, racist, and increasingly neoliberalizing society (Fikentscher 2000: 93–106). Garage was at the vanguard of this cultural identity politics, and its regulars used metaphors of family, community and church to describe their clubbing experience (ibid.: 61). In creating such an ambience in the club, the role of the legendary resident DJ, Larry Levan, was significant. Not only was Larry Levan able to create musically outstanding mixes (his musical influence in the development of dance music has been unparalleled), but he was always intimately connected to the dance floor and dexterously orchestrated its vibe, as Garage regulars have testified (Ramos 2005).

In 1987, around the time when Garage was closed down, the neighborhood where it was located—a district in SoHo close to Greenwich Village—had become a different place to that which it was when the club opened. It was no longer an abandoned neighborhood. By 1987, Garage was constantly receiving complaints from the neighborhood's residents, and on this ground, the landlord of the building refused to renew the lease. According to the 1987 zoning map, the area where Garage was located was an M1–6 zone—an as-of-right zone for cabarets—but was one block away from an R6 zone overlapping parts of NoHo and the Greenwich Village to the east and a few blocks away from the upscale loft area of SoHo. This meant that noise and pedestrian traffic from the club inevitably intersected with the lives of the neighboring residents. This exemplifies that even if a club is located in an appropriate place, administratively, it does not necessarily mean that the club can avoid neighborhood conflicts. It was also argued that the residents did not relish the fact that the dominant patrons of the club were black gays (http://www.deepattitude.com/paradise_garage.htm). The closure of Garage was also related to another factor faced by the city's beleaguered club scene, one having tragic consequences that hit clubland hard: AIDS. The owner, several staff members and the resident DJ, Larry Levan, were all dying of the disease. Not only did the disease have devastating effects on the lives and health of many in the community, AIDS also attracted popular stigma to its denizens (Musto 1987).

The closure of Garage was a huge loss for the underground club scene. For the next two years, underground clubs with a comparable quality of music, dance sensibility and community solidarity as that developed in Garage did not emerge. The club's closure forewarned that other similar underground clubs would not long endure increasing anti-dance club sentiments as well as rising rent costs. In particular, club entrepreneurs with gay black men as their primary patrons had a hard time finding places to create a club the size of Garage in gentrifying Manhattan (Berry 1992: 52). New club owners also refused to have gay black parties because these contingents, who were not economically affluent, would not bring in heavy profits (ibid.). This meant that gay blacks were losing spaces to create cultural security zones that were important for their communities. The decline of underground dance clubs also affected the operation of specialty record stores and independent recording studios, which were themselves also suffering from rising rents, neighborhood complaints about music noise, and the overall lack of interest in such music among new urban dwellers. Considering the significance of such institutions in the underground music network, this change meant that the very base of the underground scene was eroding in 1980s New York.

Such organic underground networks slowly began being replaced by the chain of expensive uptownish clubs and bars like Palladium or Nell's, evincing the gradual "gentrification of nightlife" that has given rise to the expansion of "subcultural closure" (c.f. Talbot 2006). This post-downtown termination of the underground scene would receive another blow from the planned revision of the cabaret law in 1990, which would significantly reduce legitimate spaces for dance clubs, and impose sweeping regulatory restrictions on social dancing itself.

REZONING OF THE CABARET LAW

After the *Chiasson* ruling, Mayor Ed Koch finally announced a preliminary, provisional proposal for changes in the city's cabaret zoning regulation on May 12, 1988. The process of amending the cabaret law was a two-year long process fraught with struggles between Local 802, dance/music club owners, CBs, and the CPC to forward their interests in the new zoning rule as much as possible. In this process, the municipality demonstrated its will to not lose its regulatory grip over the city's entertainment businesses, and not compromise the quality of life of neighborhoods. While the cabaret law rezoning was initiated as a response to accommodate the court's ruling over live music businesses, the municipality took this opportunity, as an "institutional fix" (Peck and Tickell 1994), to further restrict the spaces where dance clubs could be located. As described below, the city government was going to impose heavier regulations on dance venues not according to venues' sizes, but based upon whether the venue had *any size* of social dancing or not.[4] The municipality was going to impose *absolute* control over a certain subculture that had little or no popular support and, arguably, weak constitutional standing, in exchange for modestly easing regulations on live music venues. In addition, to strike a balance between protecting the quality of life for residents, and fostering musical and other entertainment in nightlife businesses and securing the city's continued status as an entertainment capital as the City Hall's official communiqués to the public (e.g. Koch 1988) frequently reminded, the solution became a spatial one—putting dance clubs in a few specific zones and zoning them out from other neighborhoods.

Approximately one month after the mayor's announcement, the *NYT* featured a story outlining the protests of residents in the Gramercy Park neighborhood against nightlife businesses (Dunlap 1988). Longtime residents and recently arrived loft dwellers in the neighborhood complained that young groups of nighttime revelers going to new restaurants and clubs in the nearby Flatiron District were causing horrendous noise, increased traffic and blaring horns at 3 or 4 a.m. on Friday and Saturday, to the detriment of the tranquility and historic nature of the area. Residents pointed to Café Iguana, Canastel's, Rascals, Stringfellows and Underground as culprits. Two policemen were assigned to the area to monitor the disorder of streets around these clubs, but residents felt that this was not enough to control the enormity of the nightlife scene in the neighborhoods. As a matter of fact, neighbors also felt absolutely helpless in improving the situation, since the zoning codes for these areas (M1–5 and M1–6) sanctioned such nighttime businesses to be located in the Flatiron District as-of-right. The *NYT* article also reported that the municipality planned a change of the cabaret zoning that would modify the Flatiron District into a zone that would require special permits for dance clubs and large nightlife businesses. Indeed, in the proposal that the CPC forwarded, the zoning code for the Flatiron District changed in this exact way.

First Proposal

In April 1989, the DCP came up with the first proposal of the zoning amendment as a new framework for the regulation of entertainment establishments in the city. This proposal divided the entertainment businesses largely into four categories. These were as follows (CPC 1989: 3–4, 6, my emphasis):

1) *Use Group 6A* included eating or drinking establishments that play accessory music for which there is no cover charge and no specified showtime. It would be as-of-right in C1, C2, C4, C5, C6, C8, M1, M2 and M3 Districts, and permitted by special permits in the C3 District.
2) *Use Group 6C* included eating or drinking establishments with entertainment (with a cover charge or showtime), but not [social] dancing and a capacity of up to 175 persons. It would be as-of-right in C4, C6, C8 and most manufacturing districts and by special permit in C1, C2, C3, C5, M1–5A and M1–5B Districts (which refers to SoHo and NoHo).
3) *Use Group 12* included eating or drinking establishments with entertainment and a capacity of more than 175 persons and/or any establishment with dancing. It would be as-of-right in C4, C6, C7, C8 and most manufacturing districts, and by special permit specified under Section 73–241 in the C2, C3, M1–5A and M1–5B Districts, provided they operated only *during certain restricted hours.*[5]
4) *Use Group 13* included eating or drinking establishments with entertainment, and a capacity of more than 175 persons, or any establishment with dancing that would operate beyond the restricted hours. It would be permitted as-of-right in C6–5, C6–6, C6–7, C6–8, C6–9,

> C7, C8 and most manufacturing districts, and by special permit under the new Section 73–244, in C2, C3, C4, C6–1, C6–2, C6–3, C6–4, M1–5A, M1–5B, M1–5M and M1–6M Districts (M1–5M and M1–6M refer to the Flatiron District).

In this proposal, the CPC created several new categories. Use Group 6A referred to establishments that used "accessory music" as background music without a cover charge and showtimes.[6] In this way, live music would be able to locate in a wider swath of Manhattan than it used to be as long as the venue would not post showtimes or demand a cover charge. Otherwise, live music would have to be located in more limited zones. Interestingly, for Use Group 6C—live music establishments that would have under 175-person capacity and would post a showtime and charge a cover charge—there was no substantial change made in the new proposal, compared to the situation before 1986, except that now the live music was regulated according to size and no longer by the number of musicians or the type of instruments.[7] The only palpable change made for this Use Group in the new proposal was that C1 and C5, which used to be off-limits to this kind of Use Group, were added as zones that would require a special permit.

Local 802 refused this proposal as it was premised on the misguided distinction between background "incidental" (or "accessory") music and other live music performances in the matter of cabaret regulation, a distinction which, in fact, had triggered the *Chiasson* lawsuit in 1986 and 1988 (Chevigny 1991: 141). The proposed amendment did not remove this distinction, but merely replaced it with different criteria—i.e. posting of showtimes and cover charge. In addition, while the new proposal seemed to relax the zoning restriction of Use Group 6C by adding C1 and C5 as possible locations for this Use Group, this was actually a cosmetic reform. Getting special permits was (and still is) extremely cumbersome and unpredictable, so changing off-limits areas to ones that require special permits was not at all useful for small venues—the very ones that *Chiasson* plaintiffs wanted to save through the litigation—that could not afford the expenses, complexity and the time consumed in the process of acquiring special permits. This is a good example of how the court ruling, even though it was concerned with the First Amendment, was not able to bring about immediate change in governmental rules. Local 802 started to mobilize popular opinion about the flaws in this proposal, and soon created allies with some CBs (ibid.: 140).

At the same time, owners of large dance clubs and some rock clubs denounced the proposal as threatening the city's nightlife businesses (Yarrow 1989).They contended that early closings would significantly impair the profitability of the clubs' operations. Regarding the changes that would require dance clubs to get special permits in a zone that used to be as-of-right to them, clubs argued that the process of acquiring the special permit would be so time-consuming (e.g. six to twelve months to secure a special permit, in addition to six months for a cabaret license) that it would discourage clubs from opening in these areas. Soon after, the until-then divided and competitive club owners formed an organization among themselves called the *New York Cabaret Association* (NYCA), in order to fight the new proposal. Interestingly, there was no mention in the NYCA's protest regarding the proposed regulation of any size of social dancing. Such a provision would predictably cause more financial distress to small clubs by imposing rules that were meant to be applied to large clubs to small clubs. Without challenging this questionable provision regarding any size of social dancing, the NYCA's protest narrowly focused on how restrictions on hours and reduction of as-of-right zones for dance clubs were going to dampen investment opportunities and the profitability of this industry. Such lacunae in the protest by the NYCA may imply that NYCA was not representative of all sizes of dance clubs in the city.

CPC planners argued that the new proposal was an effort to place dance clubs where they were most appropriate—places such as West Chelsea, which were left as-of-right for this use—and more effectively manage this use in "residential" neighborhoods. The latter included those such as the Flatiron District and TriBeCa, that would require special permits for the establishment of dance clubs (Yarrow 1989). Considering the violence and other nuisances caused by dance clubs, the new proposal made much sense, the planners argued (interview, Ken Bergen, a staff planner at the CPC, March 11, 2005). Indeed, it looked like the nightlife problems were at that time proceeding with aggravated intensity in conjunction with the popularity of crack cocaine. One *NYT* piece reported on the details of this connection and the widespread violence that was tainting clubland at that time (Freitag 1989). It cited the violence incidents, including several shootings and stabbings that led to two deaths in or near 1018 (in West Chelsea), as well as its sales of alcohol to minors and drug trafficking. It also covered similar violence in The World in Alphabet City in the East Village, Underground in Union Square, and Mars in West Chelsea. In another case, the article reported, involving the Red Zone in Clinton, neighboring residents were so frustrated with the noise from the club that they pelted its patrons and employees with eggs and bottles. According to this *NYT* article, club owners contended that of late, the police were increasingly hostile to clubs, and enforcing laws in what were previously considered to be "gray areas." The police acknowledged that the enforcement had been recently stepped up, but they cited this increase as merely a response to ballooning community complaints over violence and vandalism emanating from neighboring clubs and the increasing number of young rowdy crowds that tended to over-drink and act out.

Club owners argued that the police intentionally picked on clubs like The World, 1018 and Underground because their patrons were racially mixed teenagers (Freitag 1989). The counsel of the NYCA also sent a letter to the *NYT*, in which he argued that the *NYT* piece's description of club violence and vandalism was misleading and unfair, as it created an impression that the club scene in the city overall was tainted with violence, although a closer inspection of this article would reveal that the journalist only identified two clubs at which this was the case (Bookman 1989). He continued:

> At the root of this official harassment is a small but vocal group of neighbors with a desire to transform the commercial-manufacturing-residential districts where they live into purely residential districts, *zoning laws notwithstanding.* This, of course, ignores the mixed-use nature of these districts, into which they knowingly moved. (Bookman 1989, my emphasis)

He also argued in a public hearing of the rezoning proposal in June that the proposed amendments would have a tremendous negative impact on the number of new clubs opening in New York, especially because "[t]he hour restrictions are fundamentally anathema to this industry" (Dunlap 1989). The CPC officials at the hearing pointed out that the new proposal would not be a curfew rule, but would provide chances for operating beyond restricted hours as long as clubs acquired special permits. This answer was not convincing to the NYCA members, as obtaining special permits would be an expensive, drawn out and unpredictable process. In addition, NYCA counsel said, "few individuals will invest the considerable sums necessary for building new nightclubs while not knowing in advance whether or not a special permit will be granted and, if granted, then only for a maximum term of three years" (CPC 1989: 17). Not only would the special permit be too onerous, but it was also the case that, the NYCA asserted, the areas where clubs with full operating hours would be permitted as-of-right were too few in the proposed zoning resolution. The NYCA additionally objected to the proposed special permit provisions requiring an indoor waiting area and entrances to be located no closer than 100 feet from a residential district. While the NYCA primarily voiced its concerns about the proposal's negative impact on the *industry* and *investment*, a few other participants at the hearing also protested the new proposal on the grounds of its potential damage to the city's subcultural communities. Among those participants, a rock band called Law and Order spoke up at the hearing (Dunlap 1989). The members of the band testified that they started their musical careers in such clubs that would be substantially affected by the proposed amendments, the type of clubs in which performers such as Madonna also got their start.

While those present at the hearing all testified against the proposal, the chairperson of the CPC said that the CPC had got "thousands of letters and petitions" in favor of the modified proposal (Dunlap 1989). Indeed, most of the CBs that reviewed the proposal (as part of the ULURP process) voted overwhelmingly in favor of the new proposal (CPC 1989: 11–13). They expressed that they were willing to tolerate live music venues in their neighborhoods, and argued that discrimination against live music in favor of recorded music should be eliminated, although a couple of CBs insisted that deregulation of live music venues should not be extensive. But, they stressed that dance clubs should be subject to more restrictive zoning regulations, as social dancing venues (dance clubs) are hard to live next to *no matter what their operating hours are* (ibid.: 12). Interestingly, recurring in the reviews of various CBs of the cabaret rezoning was that in their understanding, the establishments that have social dancing were assumed to be large, and there was an important silence with respect to the regulatory approach to small dance venues. For example, CB 1 requested:

> . . . that the CPC concentrate on implementation of controls over large establishments with [social] dancing, and establish a fast-track approval process for smaller places, so that smaller non-dancing live music establishments are not discouraged by the application and licensing process. (ibid.: 12)

Here, smaller places mainly refer to non-dancing establishments. This formulation would mean that small clubs that occasionally or regularly featured dance parties be subject to the same harsh regulation and extensive limits in their location as the large discotheques if the new proposal were to be legislated.

FINAL PROPOSAL

Opposition to the initial proposal came from multiple sources. The Manhattan Borough Board, headed by jazz aficionado David Dinkins (who was elected in 1990 as the first black mayor of the city), rejected the proposal as it was applied to live music establishments (Chevigny 1991: 149). As mentioned, Local 802 and NYCA were also in opposition. The CPC's decision to change the early proposal was also made in part due to an unrelated suit brought by Limelight (a cavernous discotheque in Manhattan), in which the New York Court of Appeals decided in July 1989 that local regulations did not have the power to supersede state regulations on hours of operation of the businesses licensed by the SLA (interview, Ken Bergen, a staff planner at the CPC, March 11, 2005).[8] The CPC had to drop the restrictions on hours of operation included in the first proposal.[9]

The second proposal, thus, granted more places where Use Group 6C could be located as-of-right, at the same time as it was more restrictive in terms of places where Use Group 12A could be located as-of-right in exchange for removing restrictions on hours of operation. The proposed amendments were as follows (CPC 1989: 6–7, my emphasis):

1) *Use Group 6A* included eating or drinking establishments that play accessory music for which there is no cover charge and no specified

show time. It is as-of-right in C1, C2, C4, C5, C6, C8, M1, M2 and M3 Districts, and permitted by special permit in C3 Districts.

2) *Use Group 6C* included eating or drinking establishments with entertainment but no dancing and a capacity of up to 200 persons, and which would be permitted as-of-right in C1–5, C1–6, C1–7, C1–8, C1–9, C2–5, C2–6, C2–7, C2–8, C4, C6, C8 and most manufacturing districts. Special permits should be acquired for Use Group 6C in C1–1, C1–2, C1–3, C1–4, C2–1, C2–2, C2–3, C2–4, C3, C5, M1–5A and M1–5B Districts.
3) *Use Group 12A* included eating or drinking establishments with entertainment and a capacity of more than 200 persons, or any capacity with dancing, and would be permitted as-of-right in C6, C7, C8 and most manufacturing districts. Special permits should be granted for new businesses to open in such areas as C2, C3, C4, M1–5A, M1–5B, M1–5M, M1–6M and the Lower Manhattan Mixed Use Districts (LMM, which refers to the TriBeCa area). On the surface, only C4, M1–5M, M1–6M, and the LMM were added to the existing Special Permit requirement areas for this kind of use. However, if not Special Permit, certain other specifications would be required in C6–1, -2, -3 and -4 Districts.

The second proposal increased the capacity threshold from 175 to 200 persons as per a request from Local 802. In addition, the places where Use Group 6C would be permitted as-of-right were extended, which were not included as-of-right in the previous proposal. On the other hand, in exchange for removing restrictions on hours of operation, the CPC combined Use Group 12 and 13 in the previous proposal, and created Use Group 12A, which, compared to Use Group 12 (in the previous proposal), was more restricted in the areas of as-of-right locations. Therefore, under the new proposal, C4, M1–5M and M1–6M and LMM were no longer to be considered as-of-right zones for Use Group 12A. These included neighborhoods like the Flatiron District and TriBeCa. These four zones now required a special permit for large businesses or any capacity business permitting social dancing. In addition, a wide swath of Manhattan's commercial areas—C6–1, -2, -3, and -4—would require certain specifications before they could locate there (see Figures 5.2 and 5.3). The change that this rezoning would bring in can be measured by comparing the sizes of as-of-right and special permit zones in Figure 5.3 with those in Figure 2.1.

These particular specifications that would be applied to C6–1, -2, -3 and -4 Districts were as follows: (a) a minimum of four square feet of waiting area within the zoning lot shall be provided for each person permitted under the occupant capacity as determined by the NYC building Code. The required waiting area shall be in an enclosed lobby and shall not include space occupied by stairs, corridors or restrooms; and (b) in these districts the entrance to such use shall be a minimum of 100 feet from the nearest residential district boundary" (CPC 1989: 33). The special permit for Use Group 12A, granted by the Board of Appeal for a term not to exceed three years, required the above (a) and (b) clauses, and additionally the following conditions (ibid., 43–4), that: (c) such use will not cause undue vehicular or pedestrian congestion in the local streets; (d) such use will not impair the character or the future use or development of the surrounding residential or mixed-use neighborhoods; (e) such use will not cause the sound level in any affected conforming residential use, joint-living work quarters for artists, or loft dwelling to exceed the limits set forth in any applicable provision of the NYC Noise Control Code; (f) the application is made jointly by the owner of the building and the operators of such eating or drinking establishment; and (g) The Board [of Appeal] would also prescribe appropriate controls to minimize adverse effects on the character of the surrounding area, including, but not limited to, location of entrances and operable windows, provision of sound-lock vestibules, specification of acoustical insulation, maximum size of establishment, kinds of amplification of musical instruments or voices, shielding of flood lights, adequate screening, curb cuts, or parking. Any violation of the terms of a special permit would constitute potential grounds for its revocation.[10]

Having incorporated feedback from CBs about the first proposal, the CPC drafted more extensive and stringent terms for special permits. Responses by CBs to the second proposal were mixed, some arguing for more restrictions, say, in the form of sound-proofing measures for all entertainment establishments, large or small, and a further reduction of as-of-rights zones for Use Group 12A. Others were content with the second proposal. At the public hearing for the second proposal, Local 802 demanded any reference to accessory music should be eliminated (CPC 1989: 18). However, Local 802 stated that it agreed on regulating entertainment establishments on the basis of *patron dancing*. The effort to dissociate live music from social dancing led Local 802 to assent to the provision that would wind up doing a great disservice to one of the city's vital cultural sources—dance venues—by accepting regulations insensitive to different sizes of dance businesses. As will be seen in the next chapter, however, this strategy would backfire and wreak havoc on live music venues, too.

The NYCA counsel stated at the hearing that the NYCA objected to the expansion of the locations where a special permit would be newly required due to the difficulty of acquiring special permits. He also contended that "although existing nightclub locations would be 'grandfathered' under the proposal, the difficulty of opening a new club successfully in a location where one previously existed, coupled with the relatively short life-span of existing establishments, will result in extinction of the industry" (CPC 1989: 19–20). In addition, he requested the re-examination of the requirement that no entrance to a cabaret should be within 100 feet of a residential district—a requirement that would thwart the opening of clubs in many areas, especially in C6 zones. He further contended that the city should not require the building owner's consent in the process of renewing a special permit. In an interview with the *New York Observer*, the NYCA counsel berated the proposal as "nothing more than an attempt to take large portions of this city and make it impossible to start a nightclub there"

Figure 5.2 Zoning configurations I of Use Group 12A after the 1990 rezoning. Historical Zoning Maps used with permission of the New York City Department of City Planning.

(McHugh 1989). Before applying for a special permit, club entrepreneurs would have to sign a lease and pay rent, but there would be no way of predicting if a permit would be granted, since the club would be judged on such subjective standards as to how it would affect the future character of a residential neighborhood, he maintained.

Figure 5.3 Zoning configurations II of Use Group 12A after the 1990 rezoning. Historical Zoning Maps used with permission of the New York City Department of City Planning.

In the face of these protests, the CPC's stance toward the validity of the second proposal remained unyielding and it continued to argue for the legitimacy of the amendment (CPC 1989: 22). It insisted that the distinction between background music, which is an accessory in the restaurant and bar businesses, and other kinds of entertainment in bars and restaurants, were necessary because the CPC had observed that the former businesses had fewer land-use impacts than the latter. The NYCA's concerns were also rejected by the CPC, as the latter conceived all the requirements for dance clubs that the NYCA objected to as necessary to protect the varying degrees of the residential character and the quality of life in each zone. The CPC argued that repeated visits to dance clubs by its staff members confirmed this (ibid.: 28). The CPC also refuted the NYCA's contention that the new zoning resolution would decrease the number of new dance clubs opening in the city, arguing that there would be substantial areas in which dance clubs could locate by acquiring special permits from the Board of Standards and Appeals.

Eventually, the amended cabaret zoning law was passed on to the Board of Estimate, which in 1990 accepted it without any substantial modifications. The amended cabaret law became effective soon afterwards.

REZONING AND THE FREEDOM OF EXPRESSION

How, then, could the city administration venture to make such sweeping regulations on social dancing businesses? It appears that a hasty and unexamined assumption that dance crowds are different from regular nightlife patrons, and that once there is dancing in a venue, that this would completely alter the character of the land-use, prevailed (interview, the NYNA representative, January 14, 2005), leaving out any consideration of differentiated regulations according to the size of venues in the proposed regulatory framework.

The *Stanglin* case (1989) in which social dancing was declared as not representing a form of expression that was constitutionally protected (see the previous chapter) may have also emboldened the municipality to leave intact a sweeping zoning rule for social dancing businesses. There were cases even before *Stanglin*, in which judges opined that social dancing may not be entitled with constitutional protection. For example, in *People v. Walter* (1980), a judge ruled that the constitutionality of the cabaret law might be questionable as a violation of freedom of expression when applied to live music that involved more than three musicians, whereas, the judge mentioned in passing that it was not unconstitutional when it was applied to an establishment where patrons danced on premises. The report of this lawsuit was found in one of the internal documents of the DCA, in a document that records the history of lawsuits related to each provision of the Administrative Code of the City of New York that were related to the cabaret law.

This seems to indicate that, from earlier on, the CPC must have been aware of the fact that social dancing had a rather tenuous standing in terms of its constitutional entitlement, and it can be inferred that that was why the sweeping regulations over social dancing venues irrespective of their sizes were left uncorrected. This demonstrates how the limited scope of First Amendment entitlement can provide an avenue for municipalities to pick on certain subcultures and crack down on them when such a crackdown turns out to be convenient in governing gentrifying urban space.

In addition, in all likelihood, the municipality must have been aware that regulating social dancing in such a wholesale manner would not cause critical political volatility. There was no serious organizing going on among any segment of the affected body, such as dance club owners (at least, not until 1989) or dancers to mobilize opposition to the cabaret regulation, as the Musicians' Union had done (Chevigny 1991: 148). Indeed, there was an absence of protest about the sweeping regulation of social dancing in the nightlife sector, and as a matter of fact, it seems that hardly anyone was aware of the potential danger of such sweeping provisions in terms of freedom of expression, except Chevigny (1991: 113). The protest from the club entrepreneur organization, the NYCA, seemed more concerned with the rezoning proposal's impact on the club industry's profitability. The organization hardly broached, in public forums, social dancing as a right or as a cultural resource. While the media at the time described the NYCA as an organization that represented the city's club industry, it is likely that it only represented a certain sector of the whole club industry—the city's mega dance club owners in particular, for whom decrease in the as-of-right zones for social dancing would seriously affect their businesses, but who did not particularly care about the wholesale regulation of social dancing itself. This class character of the NYCA would linger into the 1990s and 2000s, by which time its name would be changed to the New York Nightlife Association (NYNA). In the early 2000s, NYNA would be in conflict with anti-cabaret law activist groups due to its opposition to abolishing the sweeping social dancing provision from the cabaret law. This conflict will be the subject of Chapter 7.

Meanwhile, the underground dance sector in the city seemed to be, for the most part, oblivious of the 1990 zoning changes of the cabaret law. Both in my media research and in my interviews with people who were involved in the city's 1980s club scene, I have found no evidence of political mobilization in the underground dance sector over the 1990 zoning change (interview, a former owner of underground dance club and DJ, August 23, 2006; interview, an ethnomusicologist, July 16, 2005). It may be the case that the underground scene was not aware of such changes because the effect of such change was not felt in full force until the mid-1990s when the Giuliani administration decided to take advantage of the sweeping social dancing provisions of the cabaret law to carry out a wholesale crackdown on the city's nightlife. Under these circumstances, the social dancing provisions in the cabaret law appear to have been passed without serious grassroots challenge.

CONCLUSION: THE 1990 REZONING AS INSTITUTIONAL FOUNDATION

In the 1990 rezoning text, and also in the voices of CBs, establishments with social dancing were frequently connected to the image of large and unruly discotheques, and serious discussions surrounding the value of dance clubs in the development of a variety of subcultures and in the building and nurturing of subcultural or minority communities in the city were largely absent. In addition, while clubs were frequently used to promulgate the neighborhood's image of an urban vibrancy in the real-estate market, when it came to the matter of violations of quality of life, the value of clubs that enabled such reputations was summarily dismissed. Weighed against the policy urgency and political convenience of securing quality of life and pushing forward further gentrification, it became a matter of little importance for the municipality to give serious and careful consideration to the value of nightlife entertainment, especially social dancing, and rights issues possibly encapsulated in such entertainment.

The new cabaret zoning regulation revised in 1990 would regulate social dancing not by the size of venues that would have social dancing, but by social dancing itself. This provision meant that small dance venues (or even bars that would allow a few of their patrons to dance spontaneously) would be subject to the same austere regulations that large dance clubs would be subject to, despite that the former rarely have the same degree of negative spillover impacts to surrounding neighborhoods as big dance clubs. In comments made by CBs and the CPC, small clubs represented only live music clubs, and there was no occasion in which small "dance" clubs were even mentioned in the regulatory language. In addition, social dance was recognized as not being entitled with constitutional protection, which must have emboldened the municipality to proceed with sweeping zoning regulations. The problem of this regulatory blind spot for small dance clubs has already been expressed in the previous chapter, in the complaints made by the representative of Private Cabarets Inc. in 1979, long before the 1990 rezoning, about the cabaret law's inability to distinguish between Studio 54 type mega discos and small, ordinary bars where patrons dance to music from jukeboxes. This provision, that elides regulatory codes for different sizes of dance venues, would be put into malevolent practice by the Giuliani administration in the 1990s to control nightlife businesses with a deleterious effect on people's liberty to move their own bodies together with others for the pleasure of, in particular, social interaction.

Created as a fix to solve problems stemming from 1980s gentrification, but also shaped by multiple other factors, the 1990 rezoning of the cabaret law would work as a regulatory and institutional foundation for the city's crackdown on nightlife in the next two decades. In the chapters that follow, I will examine the impact of the cabaret law's new zoning rules in relation to on-going local conflicts over dance clubs and other nightlife businesses over the1990s and 2000s, and the subsequent emergence of new and different pro-nightlife formations.

6 Disciplining Nightlife
1990–2002

The 1990 rezoning of the cabaret law drastically reduced the number of places where dance clubs could locate as-of-right, and tightened the terms of club operation in certain areas. Conflicts over the operation of nightlife businesses did not diminish, however; in fact, they continued to escalate. The reasons for this were multiple. First, gentrification intensified and expanded in the city. Second, new rules contained in the 1990 rezoning were not applied to pre-existing (grandfathered) dance clubs, which continued to cause trouble, in residential neighborhoods. These included a series of fatal incidents involving transactions of crack cocaine. On the other hand, dance clubs started to cluster in a few remaining as-of-right and affordable neighborhoods, especially in the West Chelsea area (located west of Chelsea in Figure 5.3). However, as West Chelsea also started to experience gentrification, this neighborhood, too, started to witness the same kinds of tussles between (new) residents and nightclubs that had been plaguing other neighborhoods. In the 1990s Manhattan also saw an acute increase in conflict over nuisance caused by nightlife businesses other than dance clubs, such as bars, restaurants and lounges. These businesses tend to move into, and cluster in, gentrifying neighborhoods as gentrification usually ushers in potential patrons for these businesses.[1] At the same time, despite opposition on behalf of residential communities to the expanding night scene in their neighborhoods, the State Liquor Authority (SLA) had been generous in approving liquor license applications for the purposes of urban growth (Ocejo 2009: 9). The night scene also witnessed an increasing presence of businesses (bars and clubs) that offered mass alcohol consumption for middle-class college crowds in their 20s seeking "nights out" with affordable drinks (ibid.: 123), which resulted in over-drinking among these demographics, as well as under-age drinking.

The institutional solutions to such conflicts were pursued both at the municipal and state level by residential communities. At the municipal level, in response to persistent petitions by residential communities that were increasingly organized through anti-nightlife campaigns, the municipality tried to amend the cabaret zoning regulations to de-concentrate dance clubs from mixed-use neighborhoods. They also codified the responsibilities of dance club owners, and created a multi-agency task force that was to more effectively regulate both dance clubs and other nightlife businesses. This multi-agency task force, from the mid-1990s onwards, meticulously enforced the social dancing provision of the cabaret law, which sparked outrage among nightlife actors over the legitimacy of the municipality's nightlife regulations. At the state level, residents appealed to the State Liquor Agency (SLA) to close down troublesome dance clubs and bars that could not be closed by the cabaret law, and also succeeded in reinforcing the power of Community Boards (CBs) in influencing the SLA's decision to grant liquor licenses to businesses. The SLA also passed a bill regulating the number of licensed businesses that could operate in any 500-square-foot area.

This chapter is an elaboration of how anti-nightlife governmentality developed in the 1990s through "legal complexes" (Rose and Valverde 1998)—complexes in which laws, institutions, norms and discourses work together, although they are always contested over, to produce a particular social process and geography. In this chapter, I focus on three aspects that can further explain this development. First, I show the ways in which the Happy Land fire in 1990—a disastrous fire that occurred due to the lack of proper safety measures—was frequently evoked by city officials in legitimizing tough enforcement of law on dance establishments, even in response to very minor violations that may only have a tenuous relation to safety issues.

Second, I show how the discourse of "good vs. bad neighbors" was adopted to identify un/desirable nightlife establishments, and used to penalize the ordinary features of nightlife establishments. Nightlife should be noisy and disorderly in order to provide the kind of socialization that is unique to it, and which is required to nurture creative, transgressive and counter-cultural subcultures. The story in this chapter shows how during the "quality of life" era, and in the midst of gentrification, these ordinary features of nightlife were not only decried as nuisances, but also came to be punished as *crimes*. I provide here a few examples that demonstrate the process through which the criminalization of mundane nightlife activities left only affluent nightlife entrepreneurs—often ones that stage gentrified and/or corporate forms of nightlife—as nightlife survivors, resulting in changes in the cultural geography of the city. As nightlife was needed to gain a "cool" status for a neighborhood, but then was not wanted in fully gentrifying neighborhoods, noisy, disorderly and sleazy nightlife got licensed and zoned out of these neighborhoods, whereas orderly and gentrified forms of nightlife were allowed to stay. On the other hand, anti-nightlife governmentality operated in a self-propelling manner, too, i.e. through the absorption by nightlife entrepreneurs of "good neighbor" norms, self-actualizing the latter by bowdlerizing out their wilder elements (often associated with the subcultures of ethnic/racial minority groups). As Ocejo (2009: 195–96) argues, this amounts to a downloading of policing accountability to individual owners by state and municipal governments, which had contributed to bringing bars and gentries together into neighborhoods through their promotion of nightlife in gentrifying neighborhoods, and had subsequently lost the capacity for controlling nightlife in some neighborhoods that nightlife businesses had clustered within.

Third and most centrally, I show how new laws, initiatives and institutions deployed in a crusade against nightlife, especially those implemented under the Giuliani administration, was related to Quality of Life and Zero Tolerance policing and popular "revanchism" (Smith 1996). In particular, the cabaret law regulation, the implementation of which marked a total dismissal of basic civil liberties and other fundamental rights associated with social dancing, was not an isolated incident related merely to the eccentric characteristics of Giuliani himself, but was in line with the routine dismissal of basic liberal democratic and social citizenship that has characterized neoliberal and post-industrial regimes (Brown 2003; MacLeod 2002). As explained in Chapter 1, urban space under these regimes has witnessed a series of penal policing enforced against socially and economically marginalized "undesirable" populations and land-uses that threatened to defile the image of gentrifying areas, as well as developments of commercial/residential fortresses that have served to protect middle-upper-class gentries from these "undesirable" populations and land-uses. New York City, especially under the Giuliani administration, not only took on these characteristics, but has become an uber-model for them that other cities seek to emulate.

NEOLIBERALIZATION AND GENTRIFICATION IN THE 1990S

In the early 1990s, there were signs of what some commentators called "de-gentrification," (Bagli 1991), a trend of receding gentrification, especially in the Lower East Side and the Upper West Side, following the stock market crash of 1987. This first serious stock market crash of the neoliberal era was caused by frenzied overinvestment in both residential and office/commercial real-estate markets (Smith 1998: 2–3). Signs of "de-gentrification" faded, however, during the subsequent revival of the real-estate market, which began roughly in 1994, following the recovery of Wall Street and of the economy overall. According to Hackworth and Smith (2001: 468), the real-estate market boom from 1994 onward (identifying them as "third wave gentrification") was characterized by more assertive intervention and assistance from the city government than that which took place during the 1980s, especially in NYC. This process was stimulated by reductions in federal funds flowing to localities in the 1990s, which increased the need for municipalities to generate taxes and income revenues from other sources, such as property taxes, and which also magnified the dependency of municipalities on private capital. To better appeal to private capital and the lending community, municipalities were pressured to be more market-friendly and to actively deregulate business operations (ibid.: 469–70). With the deregulation of real-estate as well as financial markets, city governments defined their roles more and more in terms of assisting profit-oriented businesses, such as private real-estate developments (ibid.: 465).

Gentrification in this period in NYC intensified in places like TriBeCa, and extended to places within the Reinvested Core (RC, see Figure 4.1), that had hitherto not been subject to full stream gentrification. These places included the Lower East Side, Clinton, the Upper West Side's Manhattan Valley and West Chelsea (Hackworth 2008: 105–42). At the same time, the geographical scope of gentrification spanned beyond the RC, reaching remote areas in Brooklyn and upper Manhattan (ibid.: 142–48). This geographical expansion to previously "risky" neighborhoods was due to the growing saturation of the real-estate market in RC areas that were highly accessibility to social and cultural amenities. Such geographical expansion to riskier neighborhoods was owed to the state's willingness to share risks with private developers that ventured into these areas (Hackworth and Smith 2001: 469). Along with the market deregulation of urban development and the rise of market-oriented governance came the militarization of urban life and urban space. This is another way that the state demonstrated its willingness to take up risks and responsibilities to cleanse gentrifying neighborhoods of "undesirable" populations and land-uses that contrasted against the image of safe and clean neighborhoods, for the benefit of developers and prospective property owners. Punitive and disciplinary policing reached an outrageous level during the Giuliani administration, including in the domain of nightlife, as I detail later in this chapter.

The "gentrification with and against nightlife" continued in this period, too. In a story that was featured in the *New York Times* (Salkin 2007), a developer *cum* landlord of several buildings in the Lower East Side was interviewed during which he reflected on his effort in the mid-1990s to promote about eighteen vacant storefronts on Orchard Street (in the Lower East Side) to establish nightlife businesses there. He rented these storefronts to bars, restaurants, clubs and experimental performance spaces which would "bring in the hipsters and change the neighborhood." Indeed, within a decade, this neighborhood, that had previously been considered a risky area for developers to invest in, went through gentrification—the change that he anticipated. He and other landlords like him were now regarded to be responsible for the "hipification" of the neighborhood that has brought with it a wave of hotels, condominium towers, and boutiques for new yuppie residents. By helping to create an artsy and hip ambiance through the encouragement of nightlife businesses, landlords like him have created fertile ground for the recent real-estate development boom. This shows that landlords and developers continued to conceive of nightlife businesses as useful, if transient, anchors for future investment in derelict neighborhoods.

As gentrification settles in, however, nightlife businesses in this neighborhood have been lambasted as the number one enemy of quality of life, as I detail in this chapter. The dilemmas of the "nightlife fix" continued.

Mayor David Dinkins (in office from 1990–1993)—a liberal Democrat and once a member of Democratic Socialists of America—has been cast as having been more sympathetic to poverty-related crimes and generous to a liberal ethos than his successor, mayor Giuliani (Gotham Gazette 2005). But Dinkins also pushed for postindustrial and neoliberal economic policies, and the promotion of gentrification was an important policy platform for him, too. Like former mayor Koch, he negotiated with firms to give them tax and regulatory breaks, such as zoning variances, to keep them from leaving the city (Fainstein 1994). He also waged a wholesale campaign against homeless people, the most well-known rendition of it being the clampdown on homeless people, squatters and pro-homeless demonstrators in Tompkins Square Park (Smith 1996). Considering that the early 1990s produced a massive number of the homeless and the precariously housed due to the economic depression, the oppression of homeless people in public space without providing proper a shelter system and affordable housings was reactionary.

The policies of the Dinkins term inevitably answered to popular "revanchism," a surge of fear and anger mostly by middle-upper classes (but also quite often internalized among lower income classes) against economically marginalized and socially excluded "undesirables," and also against liberal social policies that still remained, if to a limited degree, up to the end of the 1980s (Smith 1996). Projected in various TV dramas and shows, and movies that described racial/ethnic ghettos and the violence of people living there in a manner that resonated with the "culture of poverty" thesis, revanchism can be traced to the sense of uncertainty and insecurity caused by the recession of the early 1990s, with propertied classes blaming the victims of recession as culprits, and viciously seeking to defend their privilege. Mayor Dinkins's policies reflected this ugly politics of revanchism—shunning not only away from structural explanations of poverty, but even from any simple sense of compassion for the poor.

Dinkins was less militant with regards to nightlife than Koch or Giuliani, but there was nonetheless no real break in the policing of nightlife during his administration. At the neighborhood level, residents' complaints over increased incidents of gun violence as well as the chronic problem of noise from nightlife businesses sharply increased during this time. The furious reaction to nightlife often took up revanchist themes, vocabularies, and tones. City officials and elected politicians responded with a hardline approach against nightlife, which we will examine soon. Before this, however, I want to discuss a fire that occurred in 1990, the sweeping regulatory, administrative and police response to which would be an important bellwether of how the city dealt with nightlife throughout the 1990s and 2000s.

THE HAPPY LAND FIRE

In March 1990, a fatal fire broke out in a small, 150-person capacity social club called Happy Land, located in the Bronx. The fire was the result of an act of arson by a man who was angry with his lover, who worked at the club. The fire obstructed the only exit out of the club, and resulted in eighty-seven deaths. During the investigation of the fire, it was disclosed that the club had already been cited for violations several times, and was ordered closed. "Social club" was a not-for-profit corporation for "the purpose of providing for members entertainment, sport, recreation and amusement of all kinds" (Barbanel 1990). Once an establishment was granted the status of social club, the club was able to implement more flexible operating hours (than other nightlife businesses) and charge a membership fee. Social clubs, however, were not allowed to charge an admission fee, sell liquor or be open to the public. Happy Land, while holding a social club license, violated these rules, operating as a commercial business, selling liquor to the public, serving liquor to minors (many victims of the fire were less than 21 years old), and not being equipped with proper safety measures. It was also disclosed that the landlord did not stop the operation, despite evidence surfacing showing that the landlord knew about the illegal operations of the club.

The immediate municipal response to the fire was the establishment of the Social Club Task Force, comprised of police officers, fire inspectors, and building inspectors that would patrol and monitor for illegal operations in social clubs throughout the city. The Task Force closed thirty-six social clubs found to be in violation of fire safety laws during the first ten days after the Happy Land fire (Strom 1990). As the Task Force continued to inspect and close down social clubs with uncompromising speed and intensity, Hispanic communities, especially those in the Lower East Side and Harlem, petitioned to the Dinkins administration not to close down social clubs that served vital cultural, economic and educational needs, unlike the types represented by Happy Land (ibid.). They contended that most of these clubs, located in derelict buildings, were not in egregious violation of city codes, but would be easily shut down because they were usually housed in buildings that did not match the specific building code required for social clubs, i.e. lacking the public assembly code. Upgrading to comply with the building codes and other relevant requirements would be financially overwhelming for small, under-resourced neighborhood groups, they argued. Like artists who squatted in SoHo and NoHo areas in the 1960s, these communities were using buildings in derelict neighborhoods that were not designated for the kind of land-use of social clubs.

The repercussions of stepped-up inspections were also felt by underground dance clubs and parties that had operated as social clubs. Some of them preferred operating the venue as social clubs rather than as cabarets, because this enabled them to run their parties until noon the next day, which was an important element in completing the whole cycle of the atmosphere that these underground parties were aiming to create (Berry 1992). Legendary underground parties like Loft and the Paradise Garage (see Chapters 3 and 4) were also run as social clubs. Choice, a party that was run by Richard Vasquez in the same venue as Loft—during Loft creator David Mancuso's absence since 1988—was also a social club, and as soon as the Happy Land fire occurred, was closed down by the Task Force. Choice was equipped with proper safety measures but the building that Choice was located in was not designed for public assembly (interview, Richard Vasquez, personal communication, August 23, 2006).

At the end of June, 1990—thirteen weeks after the establishment of the Social Club Task Force—the Dinkins' administration downsized the Task Force in half, and the crackdown on illegal clubs by the Task Force became less intense. However, the powerful image of the tragic Happy Land fire and the effectiveness of the multi-agency institutional set-up of the Social Club Task Force established in the wake of the fire, nonetheless, left an important precedent in the history of governing nightlife. The multi-agency Task Force would be reborn by the Giuliani administration in 1997 as a tool to monitor nightlife, described in the latter half of this chapter. Furthermore, the Happy Land fire would become one of the most frequently recurring images about the city's nightlife, in particular, the city's dance club scene. As Chevigny (1991: 82) pointed out, this type of fire often reinforces an idea that patrons in a nightspot of *entertainment* are less aware of their surroundings due to their being fixated attention on the entertainment, and therefore, that these places warrant overly sweeping regulations. While this idea has hardly been proven as well-founded, it has continued to work to frame the authorities' approach towards, and justify, any stringent regulation of social dancing businesses through an application of very meticulous provisions.

EMERGENCE OF NEW LAWS

The 1990 rezoning of the cabaret law created a legal system through which the city government could restrict the places where new social dancing businesses were allowed to locate as-of-right. However, nightlife conflicts did not discontinue. The following details why and also how the municipality as well as the state government responded with new laws.

Changes in the Cabaret Rules

In August 1990, Mark Green, the Commissioner of the Department of Consumer Affairs (DCA) announced a set of proposed rules. The commissioner continued, "While many well-intentioned people go to clubs, some of the worst clubs are breeding grounds for violence, crime and noise" (Nieves 1990). Commissioner Green announced further that twenty clubs were responsible for the bulk of the 1,500 complaints that the DCA had received in the previous six months. He even mentioned that some clubs attracted patrons who "gathered outside with their own 'boom box' entertainment, as well as knives and guns" (Glaberson 1990), thereby making a putative nexus between black and Latino subcultures and violence. As a symbolic gesture, Green delivered this announcement in front of the Red Zone, in the Clinton neighborhood, a club which accounted for 10 percent of all the complaints. The vice president of the 53rd–54th Street Block Association said that, "Since this club opened, I've noticed an exodus of people I've known for fifteen years" (Nieves 1990). The club was noisy and the crowds were intoxicated vandals, the president added. To the dismay of the Association, however, there were no grounds on which to close the club according to cabaret law provisions, as under the cabaret law, businesses were not penalized for only generating excessive noise, crowding or vandalism, as long as they were located in a proper zone and equipped with proper safety and other requirements. This was the reason why communities and the Association started to appeal to the SLA to close down noisy and raucous clubs.

In response, the owner of Red Zone contended that residents unfairly blamed his club for bringing in crimes and nuisances, because Hell's Kitchen (within the Clinton neighborhood), where the club was located, had been a problem area long before Red Zone opened. For residential communities, however, the club was a symbol and a spawning ground for the criminality of Hell's Kitchen. It was also reported (ibid.) that the owner of Red Zone was involved in the secret ownership of Underground in Union Square—another club also included in the commissioner's list of troublemaker clubs. Once found to have made a false statement about the ownership, the Underground's liquor license was confiscated by the SLA in 1990. Major newspaper reports on club violence often accompanied stories like this, stories about club owners' clandestine ownership of other clubs or suspicious club management practices, such as tax avoidance. These reports, certainly, were true of some dance clubs, but they also reinforced, in their wide circulation, a popular perception that club owners were affiliated with the underworld and with black market money, if to varying degrees. This representation tended to generalize and reinforce the image of dance clubs as profit-driven enterprises, underplaying the roles as cultural institutions that many underground/alternative clubs were playing. Such representations often buttressed popular endorsements of sweeping regulations of dance clubs.

Apart from the ownership scandal, Underground had already had numerous other problems, including incidents of fatalities.[2] Calling the Underground a "magnet for violence and an impediment to their efforts to rebuild the neighborhood" (Freitag 1990), a coalition of Union Square residents, shopkeepers, community leaders, politicians and universities (New York University and the New School for Social Research) waged a war against the club. The leader of the coalition (and also the executive director of the 14th Street-Union Square Local Development Corporation) voiced his concern that the violence from the club would negatively affect the revitalization of the neighborhood, scaring away people who might otherwise consider moving into the Union Square and neighboring Flatiron areas, both of which had seen the explosive revival of their economies with the development of refurbished upscale lofts and condos (ibid.). In February 1989, responding to the coalition's plea, the SLA charged Underground for serving alcohol to minors and for being a focal point of police activity, but the case was repeatedly delayed. A spokesman for the SLA responded to the enraged coalition by arguing that the club was entitled to due process regarding a charge that the club argued was unfair. The club argued that most of the crimes it was charged with occurred outside the club, for which the club should not be held responsible. They argued that the club had been making efforts to improve matters: bouncers were always stationed at the door and inside, and all patrons were frisked before being allowed entry.

The club started to work with residential neighbors in order to appease them, starting to negotiate the terms of operation and the content of club parties. In a meeting with the coalition, the club's managers promised that they were going to hire uniformed guards to patrol the area outside the club. In addition, the club vowed that it would make changes to the content of parties hosted in the club, and the marketing strategies of these parties, in order to "draw an older, more affluent crowd" (Freitag 1990), while at the same time getting rid of the reggae music parties on Friday nights that attracted younger patrons. Soon after, the managers changed the club's name from Underground to Palace de Beauté. According to the spokeswoman of the club, contrary to the rougher crowd that the Underground used to attract, the new club was "upscale" and "regularly filled with 'young hip kids' and models" (Glaberson 1990). It is important to notice that the solution to eliminate violence was sought by up-scaling and whitening the club's crowds—a phenomenon that would become more common in the club world during the 1990s. This shows how gentrification and the subsequent neighborhood conflicts over nightlife has led to what Talbot (2006: 159) calls "subcultural closure," as the process forced the nightlife scene to purge wilder parts of itself, the parts more associated with zany but creative and experimental subcultures often associated with racial/ethnic minority communities, and transforming themselves into more gentrified institutions that catered to upscale boulevardiers. This is one crucial way that anti-nightlife governmentality has functioned in the city.

In response to the increasing problems related to dance clubs, new rules were proposed by the DCA commissioner Green that would codify business owners' responsibilities for maintaining order outside as well as inside clubs (Glaberson 1990). More specifically, owners would be required to make sure that their patrons would not crowd outside their clubs, creating noise, blocking sidewalks, or causing any other disturbances in the area.

They would also be prohibited from hiring as bouncers anyone with a record of conviction for a violent felony. The most remarkable provision in the commissioner's proposal, however, was that from then on, new cabaret license applications were to be forwarded to Community Boards (CBs), which, then, would be given forty-five days to provide a non-binding resolution about the applicant to the DCA (Nieves 1990). CBs were already allowed to advance their opinions about granting special permits to clubs by the Board of Standards and Appeals. CBs were now provided with another institutional weapon through which they could voice their views for or against cabaret applications *even* in as-of-right zones. These new rules were codified into the *Rules of the City of New York*, and took effect in September of 1990.

In response to the commissioner's proposal, the counsel of the New York Cabaret Association (NYCA) asserted that the commissioner and other officials were making a sweeping generalization about problems that, he argued, were limited to some clubs—that happened to be non-members of the NYCA—that were not being operated by skilled professionals (Glaberson 1990). As will be revisited in the next chapter, organized nightlife associations sought to convey a public image of themselves as professional entrepreneurs that deliver hospitality services to consumers, and distance themselves from the characteristics of club owners from the past, who were seen to have operated at the surface of a deeper black market.

Problems in West Chelsea

By the early 1990s Chelsea had become "Manhattan's disco district" (Howe 1992a), as this area, especially West Chelsea, was one of the few places in Manhattan that had not yet been (re)developed in earnest (so rents were relatively cheap), and was also one of the few areas left as-of-right for social dancing businesses after the 1990 rezoning of the cabaret law. The mega discos, small dance clubs and bars in West Chelsea were housed in warehouses, garages, and abandoned manufacturing plants scattered in the area west of 10th Avenue between 14th Street and 34th Street. Even though these clubs were located far west of the residential districts in Chelsea, this did not mean that major clashes between residents and nightlife businesses could be avoided. Since many of the nightlife patrons used public transportation, these patrons (who numbered in the thousands) had to pass by residential districts to get to subway stations at the wee hours. Residents were enraged by nightlife patrons who "are boisterous, fight, set off car alarms, trample gardens, litter sidewalks and sometimes even urinate in hallways" (ibid.).

As complaints and petitions over the operation of dance clubs increased, CB4's Disco Task Force and State Assemblyman Richard Gottfried organized a public hearing. At the hearing, the NYCA representative contended that clubs were being unfairly accused:

> Clubs cannot control what happens in the streets . . . We think we do remarkably well in providing a safe and fun environment for New Yorkers and tourists, given the hours we're open and the large numbers of young people we serve. (ibid.)

Club owners also came out to plead that clubs had been doing their best in their efforts to control crime, vandalism, and public nuisances. The manager of the Sound Factory in the Flatiron District—known as the successor of the legendary club, Paradise Garage—attested that his club's patrons were mostly mature gays, and that they had been denying entry to aggressive young people. The manager of Roxy said that the club was doing its best to sustain a good relationship with the community; that is, the manager frequently attended the CB's meetings, and that the club had installed extra lighting, cleaned the surrounding streets regularly, and had put a sign outside the club asking patrons to be orderly and considerate of their neighbors.

Chelsea residents did not agree. Residents drew attention to several incidents of stabbing, knifing and mugging in the clubland area of the neighborhood. The commanding officer of the 10th Precinct commented at the hearing that the nightlife problem should be addressed more structurally by changing the zoning designation of the area, rather than by reacting to individual incidents only when circumstances demanded. Indeed, most of the clubs that were at the center of controversies were legitimate businesses, located in an as-of-right area for dance clubs. Zoning remained a thorny issue among nightlife businesses right through to the 2000s, when a series of zoning amendments were proposed for the neighborhood to usher in more residential developments (see Chapter 7). West Chelsea would not any more exist outside of the interests of real-estate capital that sought to take advantage of the booming real-estate market in Manhattan. Nightlife and zoning codes for the area became a conflict-ridden issue as these rezoning proposals proceeded in the early 2000s.

The 200-Foot Bill

In July 1992, Commissioner Green of the DCA proposed another bill. The bill proposed a modification of the zoning regulation of the cabaret law, so that medium to large-sized dance clubs with a capacity of 200 or more would be newly prohibited from opening within 200 feet of residential buildings, although existing clubs would be able to retain their licenses as long as the current owners held a controlling interest (Howe 1992b). The bill also proposed that the DCA would be given the power to shut down "disruptive" dance clubs—clubs that were extremely loud and were known to have a history of violence and/or attracted rowdy crowds. It was proposed that if a club was cited twice or more with noise convictions in a year (e.g. noise level exceeding forty-five decibels, or outside noises being so loud that it could be heard clearly in a neighboring apartment), then the DCA would revoke the cabaret license of the club.

The bill, despite even having mayor Dinkins' endorsement, however, confronted unexpected opposition in City Council, and its passage was delayed. This was in contrast to the swift move in City Council at the same time to pass an ordinance limiting locations of adult video stores and topless bars. A Councilmember who was in favor of the bill, stated that he was surprised to see that nightclubs, "which appear to have a weaker claim to First Amendment protection" than adult video stores, were harder to interdict than the latter—another sign that politicians were aware of the weak constitutional standing of social dancing. The chairperson of the Council's Community Affairs Committee was against the bill, arguing that the bill would be "a one-sided construction which puts the onus on the disco owners" (Howe 1993). Councilmember Thomas K. Duane who represented Chelsea and Clinton areas was the most fierce opponent of the bill, which he called "a wild card." He requested a land use study for the DCA-proposed 200-foot bill, which successfully delayed its passage.

Duane, then, published an op-ed in the *NYT*, beseeching the city to "Save Our Discos" (Duane 1993). He asserted that the 200-foot bill would have dire implications on the city's nightlife. Duane argued that because most commercial areas in NYC had by now become mixed-use areas (areas in which commercial land use was combined with residential ones) not in the least due to the conversion of loft spaces in commercial areas into residential uses, the proposed bill would eliminate the few remaining commercial zones that discos could legally locate in. "Such misguided anti-discotheque legislation," he pitched, "would stop New Yorkers from dancing and partying, which were important forms of socialization not only for trendy Manhattanites, but also for ethnic minorities." A coalition called Save Chelsea (founded in the summer of 1991 in order to pressure the administration to regulate large nightclubs in Chelsea) blamed him for poorly representing the very community that elected him. The former DCA commissioner Green published a counter-argument, in the *NYT*, to Duane's op-ed (Green 1993). Green argued that Duane's approach to the 200-foot bill was ungrounded and simply "echoed the hyperbole of club owners." More than half of the current licensees in the city, he argued, were either under-200 in capacity or more than 200 feet from a residence, and that therefore, the nightlife industry would not be ravaged by the 200-foot bill, as Duane feared. The bill, Green argued, was narrowly targeted to mega clubs that tended to attract large, disorderly crowds.

The 200-foot bill was eventually not passed, however, to the dismay of residential communities.

Lease Disputes

While municipal initiatives to legislate new regulatory provisions into the cabaret law were hounding the city's clubland, some individual clubs were also embroiled in disputes with their landlords. Capitalizing on the rising property values of their gentrifying neighborhoods, landlords often tried to push existing tenants out, so that they could lease the premises to higher-rent businesses. The venerable Bitter End on Bleecker Street in the West Village—a 32-year-old coffeehouse *cum* performance club, and the home of artists such as Carly Simon, Joni Mitchell and Judy Collins—was a case in point (Deutsch 1992). A building inspector visited the club twice and booked it for violations after having found that the seating and floor plan had not been posted on the wall as was required by the law. Based upon these violations, the landlord ordained the club's eviction, but also proposed to let the club continue to operate if the club agreed to pay triple the monthly rent over the next five years. The landlord also asked the club to sign a twenty-one-page stipulation of settlement which, according to the club's lawyer, would make it vulnerable to eviction even in the case of minor violations. The situation seemed simple and clear for the club; the landlord was using the violation as an excuse to raise the rent, or to attract more profitable tenants in the building by kicking out the Bitter End. The club's fans and artists soon started a campaign to save the club, signing petitions and writing letters to local politicians. Some veteran artists held benefits to help the Bitter End to pay legal costs and their increased rent. Jann Wenner, founder and publisher of *Rolling Stone*, even wrote to the City Council President, suggesting that the club be made a landmark, to which the President's office responded positively. The Bitter End was awarded landmark status by the City in 1992, and has continued to survive to the present day. While this was a celebratory achievement, it was also obvious that the same fate was unlikely for lesser-known venues. The institutional protection of music venues was not extended to obscure clubs juggling with increasingly prohibitive rents that could not exert the kind of political clout that Bitter End could exert.

Changes to SLA Laws

Residential communities had been complaining that the SLA was not as responsive to immediate local demands for closing noisy liquor-licensed businesses, and that, even worse, it had been generously issuing liquor licenses for the purposes of economic growth. They argued that this had inevitably escalated local conflicts over nightlife, but had also had the effect of displacing businesses that cater to local demands (such as laundry shops, hair salons, etc.) as bars elevated rents to levels that these businesses could not afford (Ocejo 2009: 165). However, two changes occurred in the SLA that would set the tone for a different history with regards to nightlife regulations in troubled neighborhoods. In September, 1993, state law was changed so that a CB would be able to hold a public hearing to challenge the SLA's decision to grant a liquor license to a business if three or more licenses already existed within a 500 square foot area, and the SLA would have to strongly consider the CB's objection to the license (Howe 1993).[3]

While the SLA was not obliged to accept the CB's opinion, the passing of this law became a watershed moment in the history of anti-nightlife activism by residential communities. In densely populated Manhattan, it would be hard for a new license applicant to find a spot that had fewer than three licenses within a 500 square foot area. This was a salutary achievement for CBs, as they would now have direct influence over decisions regarding the opening of new dance clubs, which they have not been able to do under the cabaret law if the area fell as-of-right for "social dance establishments."

Another law also took effect in November 1993 which mandated that the SLA should consider "public interest" factors like traffic, noise, reported criminal activity, and the history of violations on the premises in granting liquor licenses to businesses (Howe 1993). Earlier, it was usually the criminal record of business entrepreneurs or the criminal record of the premises that mattered in the granting of liquor licenses. But, through this new law, nuisances were now also counted as factors in determining the legitimacy of a business. Noisiness and disorderliness needed to be managed because they broke down quality of life. But it also seemed, in the logic of officials creating these new regulatory practices, that noise has a criminogenic nature,and therefore, curbing noise in the city will often be tantamount to governing crime; an instance, that is, of a quasi-Broken Windows thesis influencing decision making "(see Chapter 7, pp. 142–3).

The effects of the two new SLA laws were felt immediately. In the Flatiron district, residents who had been weary of problem nightspots utilized the 500-foot rule to prevent three new clubs from opening. One club owner sought to convince CB 5 that the club would cater to a gay clientele, "a nonviolent and fun-loving group" (Howe 1993), which initially seemed to convince CB members. However, the club was eventually opposed, as community members discovered that the owners once ran 1018, a disco in the 1980s notorious for numerous shootings and stabbings. In another case, CB 3 rejected an application for a new nightspot in 1995 in the area where the Sunshine Theater is located on East Houston Street (at the border of the East Village and the Lower East Side). The representative of the new nightspot contended that the club would be modeled after the Brooklyn Academy of Music (BAM), "featuring tony, ticketed events, not sloshy club and party nights" (Farber 1995). CB3 initially welcomed it, but ultimately rejected the liquor license of the club based upon the 500-foot bill, after a private party rented on the spot before the opening of the club turned obstreperous. An anonymous club manager who went through a similar experience of community opposition told the *Village Voice* (Bastone 1997), "You have to kiss [CBs'] ass from here to Times Square."

By the mid-1990s, community anti-nightlife coalitions mushroomed. In the language of these coalitions, nightlife was discreditable as profit-ravenous businesses, the cause of nuisance and an invitation for urban ills to take over the neighborhood. The hegemony of this discourse marginalized and excoriated counter-discourses articulating that nightlife had actually contributed to neighborhood revitalization, and that there is nightlife that played an important part in cultivating diverse urban subcultures and alternative politics. As Ranasinghe and Valverde (2006) pointed out, the political clout in the land-use system mostly lies with property owners, so this often makes it highly challenging for marginalized and stigmatized groups to advance their interests; such has been the case for much of the nightlife community.

However, it was observed that the SLA was to a certain degree reluctant in regulating (the number of) nightlife businesses. The SLA made frequent exceptions to the 500-foot rule, generously granting liquor licenses to applicants, especially bars. According to Ocejo (2009: 154–160), the SLA, under the State's general orientation towards post-industrialism and entrepreneurialism, tended to see liquor serving nightlife as a vehicle of economic growth as it improves the tax base, employments and images of neighborhoods. The SLA sometimes saw granting liquor licenses as falling roughly within the purview of "public interest" as a useful transient use in a blighted neighborhood as they occupied vacant commercial storefront, and made exceptions to the 500-foot rule when it issued liquor licenses in these cases.[4] Most of all, liquor serving businesses paid licensing fees to the State, as well as fines in cases of violations, and the alcoholic products that they served amount to tax resources for the State, too. The State's interests in liquor serving businesses as such inevitably led the SLA into conflicts with residential communities that interpreted "public interests" more in terms of their own quality of life.[5] Residential communities sometimes took the 500 feet rule exceptions that the SLA made to the appellate New York State Supreme Court to challenge them, the Court mostly ruling in favor of the residential communities. Residents also reviled the lack of SLA enforcement, when liquor license applicants exercised "bait and switch" tactics—when an owner applied as a restaurant in order to appease residents who preferred restaurants to bars in their neighborhoods, but then changed the business into a bar once the license was granted (ibid.: 173). The SLA had only seven investigators for the New York metropolitan area, which pointed to a structural drawback, rather than deliberate negligence, of the SLA enforcement. The SLA would be continuously pressured by CBs and city politicians to fix problems of enforcement throughout the 1990s and the 2000s.

CRACKDOWN ON NIGHTLIFE DURING THE GIULIANI ADMINISTRATION

Quality of Life Policing by Giuliani

The standoff between distressed residents and nightclubs entered a new phase with the advent of mayor Giuliani, who was elected in 1993 and again in 1997, primarily on a "quality of life" platform. The phrase, "Giuliani time" (Smith 1998), is often used to characterize the period of the militarization of urban life and urban space under mayor Giuliani in which "undesirable" and "disorderly" populations, land-uses and behaviors were unduly penalized and often displaced in order to enhance security and the quality of life in the city. The Quality of Life policing by mayors Koch and Dinkins was dwarfed by Giuliani's Quality of Life initiatives and Zero Tolerance policing, and the latter's scandalous authoritarian persona reinforced the sense of this historical break even more. The "undesirables" and "disorderly" referred not only to serious criminals, but a range of demographics: i.e. victims of neoliberal economic restructuring, such as homeless people, panhandlers, sex workers, squeegee cleaners and poor minority youth; populations causing property damage, such as graffiti artists; land-uses that defiled the image of public space, such as sex shops and street vendors; people committing misdemeanor offenses, such as drunk people in public space, jaywalkers, "reckless" bike riders and unruly cab drivers; and demographics that symbolized political liberalism, such as street protesters and public sector unions. The administration blamed these populations as the "visible signs of a city out of control" (ibid: 3), which resonated with the "Broken Windows" theory.[6]

The NYPD was granted an increased budget and unprecedented authority, the inevitable consequence of Giuliani's devotion to a law and order paradigm.[7] The mayor's office authorized the NYPD to exercise broad and flexible means to control "disorders" in public space, including authorization to stop and frisk anyone deemed suspicious.[8] The severity of the zero tolerance approach taken by the NYPD was mostly felt by people of color: blacks, Latinos, and people that looked like the immigrants. Most widely publicized was police brutality visited upon the innocent Haitian immigrant Abner Louima in 1997, the police killing in 1999 of innocent West African immigrant Amadou Diallo, who was shot forty times, and another guilt-less Haitian immigrant, Patrick Dorismond, who was also shot dead (McArdle and Erzen 2001). Homeless people were also vulnerable to police brutality, but the mayor, who sought a concerted crackdown on homeless people and the de-funding of homeless services (Smith 1996: 224–25), often granted amnesty to police perpetrators. The Coalition for the Homeless formed an organization called Streetwatch, to monitor the police's treatment of homeless people, and eventually filed a multi-million-dollar suit against the police (ibid.: 224).

Whether this pre-emptive policing indeed contributed to reducing crime rates as the mayor and the Police Commissioners claimed is still a point of dispute (Smith 1998: 4).[9] But commentators pointed to the reality that this type of policing certainly contributed to enhancing the *sense* of safety and security that the city had long lost. Not a few scholars have extolled Giuliani as the mayor who restored livability to the city (e.g. Jackson 2005). The conservative Manhattan Institute, the MAC and the state's EFCB (see Chapter 3) were also key players, and based positive evaluations of Giuliani's aggressive law enforcement on the views of rating agencies (Mitchell and Beckett 2008: 85). It was an important change, because it would mean that the city now had an enhanced reputation among the middle-upper class, corporate executives and the lending community, all of whom the city sought to appeal to. Popular revanchism against disenfranchised populations had its heyday under the Giuliani administration, as was demonstrated by the popular support for punitive policing that helped elect Giuliani to his second term as mayor in 1997. Smith (1998: 3) argues that Giuliani appropriated the prevalent insecurity and increasing revanchism in the popular mindset engendered by the early 1990s recession, and skillfully reflected it in his policy platforms in order to consolidate his electoral basis. Giuliani later gained even international credit for having established a safe city, and the "New York Model" of expertise on crime fighting has since been exported to municipal governments world-wide (cities in Mexico and South America most prominently) through the mediation of such transnational security consulting firms as Giuliani Partners L.L.C. (which Giuliani himself founded) and The Bratton Group L.L.C. (which William J. Bratton, one of the police commissioners under mayor Giuliani, founded).

The Giuliani administration sought to firmly inscribe a business-friendly image in addition to that of crime-fighting, for NYC. The administration presented itself as a neoliberal state in the "roll-out" stage (cf. Peck and Tickell 2002), actively implementing privatization, tax incentives and other financial and legal largesse to corporations. In the so-called "corporate retention deal," tax breaks, zoning incentives and other forms of subsidies were granted to stay put those corporations which threatened to relocate elsewhere, costing the city taxpayers an average of $125 million per year, and surpassing the volume of largesse handed out by Koch or Dinkins (Bagli 2005: 57). These largesse cemented the city's dependence on two of most volatile and crisis-prone industries—finance and real-estate. In contrast to these "corporate welfare" policies, mayor Giuliani took welfare benefits away from those who needed them more urgently than ever, and reinstituted benefits in the form of workfare programs. He made radical budget cuts in public spending to higher education, and trimmed government expenditures through the privatization of public service and public resources, and lay-offs to the municipal workforce (Mitchell and Beckett 2008: 85).[10] Business Improvement Districts (BIDs) had taken on a central role in the upkeep of the city's parks and other public spaces (Zukin 1995: 34–38). While these neoliberal policies sharply increased disparities between haves and have-nots and further isolated people of color, mayor Giuliani justified his policies, underscoring how his neoliberal policies led to positive bond-ratings for the city (Mitchell and Beckett 2008: 88).

The disintegration of social policies, outright support for neoliberal rationality and private capital, and punitive urban policing against the economically and socially marginalized went hand in hand with the quotidian suspension of democracy by the municipality. The Giuliani administration took after the neoliberal state that Brown (2003) describes as routinely disdaining civic liberties, even those enshrined as the basic tenets of liberalism, such as free speech and free assembly. Needless to say, the Giuliani administration had to confront lawsuit after lawsuit, losing more than twenty-five involving the municipality's violations of the First Amendment (Newfield 2002: 16). The municipal decision to de-fund the Brooklyn Museum of Art for what the mayor branded as an anti-Catholic exhibit, its retaliation against Housing Works, a militant AIDS advocacy group in the form of the withdrawal of funding for being publicly critical of the mayor, and the banning or regulating of the size of rallies around City Hall are exemplars of these violations. Counter-activism notwithstanding, Giuliani's economic, social and penal policies had purchase with broad segments of New York's residents, bespeaking the gradual transformation of popular perceptions of justice and democracy and the reconstitution of the geography of them, accordingly.

Three People Dancing

The Giuliani administration proclaimed that it would seek to claim back the city's public space in the document "Police Strategy No. 5," published in 1994. To achieve this goal, the City first sought to purge the city of homeless people. This was followed by a proposal for a new restrictive zoning regulation for "adult establishments." The City contended that these establishments caused "secondary effects"—increased crime, decline in property values, negative impacts on other, "legitimate," commerce, and other quality of life violations. The move to zone out adult establishments generated protests by various groups. Queer activists contested the proposed zoning changes, voicing that the zoning regulations would disproportionately zone out gay spaces such as bars, bookstores and clubs from traditionally gay neighborhoods, such as Christopher Street, the West Village and Chelsea (Buckland 2002: 130). Connecting these zoning changes to the revanchist movement, others (e.g. Papayanis 2000) also maintained that the municipality's concerted efforts to gentrify urban space and attend to the concerns of the middle-upper class had been trumping any serious consideration of civil liberties of free expression, driving out people and activities that were glossed as representing anti-family and sleazy subcultures, and resulting in the suburbanization of urban life. Concerned groups took the proposed zoning regulation of adult establishments to court, but did not succeed in convincing the court of the unconstitutionality of the zoning regulation.

It must not have surprised anyone when mayor Giuliani started, thereafter, turn his attention to nightlife businesses as enemies of the quality of life in the city. Dance clubs, like Sound Factory in the Flatiron District, one of the most popular underground dance clubs at that time, were relentlessly pursued with drug and other criminal charges. Other than heavy criminal charges, however, smaller infractions of existing laws or accusations of nuisance-causing were also severely punished by the police to keep clubs in line. In this policing a multi-agency task force (henceforth Task Force) created in 1995 comprising city agents from each department related to nightlife inspection, including the FDNY, NYPD, DOB, DCA was central (ibid.). This Task Force would make surprise visits to clubs and fine or padlock the club if it was found to be breaking the law. It took after the Social Club Task Force created by former mayor Dinkins, but this time, it aimed to police legitimate as well as illegal businesses, contrary to the patterns of enforcement under mayor Dinkins (interview, NYNA representative, personal communication, January 14, 2005).

Quick and powerful effects ensued. The Task Force issued numerous tickets on their visits to businesses. Sometimes, the penalty was imposed through reasonable readings of the law, but on many other occasions, penalties were based upon nitpicky interpretations. This sweeping enforcement stirred business owners to conjecture that the crackdown was being carried out under disingenuous pretenses of maintaining order and safety. It was, they suspected, intended to harass nightlife businesses in order to make it harder for them to operate in the city (interview, NYNA representative, January 14, 2005). This was particularly because such harassments included the penalization of nightlife businesses that did not have a cabaret license. The Giuliani administration interpreted social dancing as consisting of three people dancing together, and the Task Force started to police compliance of this dictum strictly. If more than three people were discovered moving rhythmically together in establishments not licensed as cabarets, it would now constitute a violation and would be subject to the issuance of a ticket by the Task Force. As a fairly large portion of the commercial area where people could drink was not zoned for *any size* of social dancing after the CPC's rewriting of the cabaret zoning law in 1990 (see Figure 5.3), any casual dancing by more than three people that could happen in any ordinary club, bar, or even restaurant, could become a violation.

From late 1995 through May 1996, the Task Force, now derided as "dance police" by nightlife actors, issued 10 citations and closed two bars for operating as dance clubs without proper cabaret licenses (LeDuff 1996). Bars and clubs in the rapidly gentrifying East Village and Lower East Side were particularly hard hit by this crackdown because there were few streets in this neighborhood that were as-of-right for social dancing. Den of Thieves, a "progressive bar" on the Lower East Side, was also cited for illegal dancing in 1995 by the Task Force. The owner paid no more than $60 as a fine, but was told that one more violation would involve the club being padlocked. Grumbling that he redesigned his venue to discourage patron dancing, he continued, "People dance at Taco Bell if there's a good tune on" (Owen 1997a). Nation, and Lakeside Lounge in the East Village and Club 205 on the Lower East Side were also cited for allowing patrons to dance (Katz 2000). Bars and clubs put up a sign that read "No Dancing" (see Figure 6.1) People started to analogize this situation to the

Kevin Bacon movie from the 1980s, *Footloose* (interview, LDNYC leader A, personal communication, July 29, 2002).

The DCA spokesman stated that the Task Force merely worked in response to complaints being made by residents (LeDuff 1996). What was controversial to such a seemingly straightforward ground for enforcement was that resident complaints were mostly about noise, congestion and rowdyism outside nightlife venues, but not about a few people dancing inside the businesses. Once the Task Force got calls from residents, however, it went to the premises and inspected it thoroughly based on a range of laws applying to nightlife businesses, including the cabaret law, that had nothing to do with the subjects of the complaints. A spokesman for the Task Force responded to a question about the three people dance regulation by saying, "What we are trying to do is avoid another Happy Land" (ibid.), evincing again how the symbolism of the Happy Land fire—of righteous concern about safety—was utilized to explain the meticulous regulation of dancing and to pre-empt resistance to meticulous cabaret law enforcement. The Happy Land fire produced a calamity of such a scale because the club was not equipped with the safety measures that a business of its size should be equipped with, so equating of Happy Land to bars and clubs that allow a few patrons dancing while actually implementing safety measures proper to its size, seemed inaccurate and unfair. Despite this flaw of reasoning,

Figure 6.1 "No Dancing" sign in Plant Bar. Source: http://www.thirteen.org/nyvoices/features/license.html.

this equation, which was conjured up repeatedly among governmental officials, planners and judges, effectively normalized popular perceptions of the volatility intrinsic to nightlife, and performed as part of the authority of expertise that justified strict and meticulous regulations.

Crackdowns based upon the three-people-dancing provision were a clear reminder to nightlife businesses about the degree of power (and abuse thereof) that the state can wield to enforce control. There was a sense among owners that the City did not want them around, and wanted them to move out of Manhattan (Goldstein 1997) in order to make expensive quarters like the East Village and the Lower East Side more amiable to the quality of life demands of newcomers. The owner of Dens of Thieves opined that the dance police's crackdown had much to do with the recent boom in the real-estate market in the Lower East Side (Owen 1997a). Ironically, it was this neighborhood where landlords had sought to revitalize the area by having nightlife businesses as tenants at the storefronts.

On the other hand, this period also saw the increase in "lounges," a type of nightlife business which had the feel of dance clubs, with DJs and often small dance floors, but which did not have cabaret licenses. This was a tactic developed among nightlife business owners in response to the restriction of zones where any size of businesses with social dancing could be located in Manhattan (interview, NYNA representative, personal communication, January 14, 2005). The authority's response in the form of the three-people-dance rule enforcement of the cabaret law, therefore, could be conceived of as legitimate. However, such a tactic developed by owners simply betokened the fundamental flaw of the cabaret law and its adverse repercussion—that is, the cabaret law's restriction of *any* size of nightlife business that had *any* size of social dancing to very limited areas of the city no matter how different these businesses' nuisance effects may have been, depending, for one, on the size of the business. This structural flaw remained unquestioned in the authority's drive to enforce the three-people-dance rule. I revisit this issue again in Chapter 8, where I discuss the judicial process in which this issue was debated between anti-cabaret law activists and the City.

Continuing Crackdown On Nightlife Businesses And the Formation of the NYNA

The three-people-dancing regulation was merely a part of a much broader crackdown on nightlife. The emerging rave scene, led by DJ Frankie Bones and Adam X, based mostly in the Bronx and in Brooklyn, was ruthlessly closed down by the Task Force (interview, DJ Adam X, personal communication, May 18, 2005). Nightlife businesses were also fined and padlocked for variegated reasons; one nightlife veteran whose premises was once visited by the Task Force calling this regulatory practice "creative ticketing" (Bastone 1997). For example, Wetlands Preserve (henceforth, Wetlands) in TriBeCa was cited for violations because bands put up illegal posters advertising their appearance at the club (Bumiller 1996). In response to criticism for this type of law enforcement, Mayor Giuliani defended his efforts to rein in a "menace" (ibid.) as part of his broader campaign to secure quality of life in the city. Whether or not clubs like Wetlands were a "menace" is highly disputable. Wetlands opened in 1989, and since then had been a reputable venue for environmental activism and the home of the burgeoning jam band scene. The club's revenues were used to support activism, and the club even successfully developed a campaign to persuade large corporations, such as Home Depot, to cease the sale of wood from environmentally sensitive areas (World Beat 2001). The club also hosted artists such as Pearl Jam, Sublime, Travis, David Gray, Counting Crows, Oasis, and Rage Against the Machine in their early careers, and invited artists from a wide range of other musical genres.

In response to escalating confrontations between residents and clubs, David Hershkovitz, the editor and publisher of the downtown entertainment magazine, *Paper*, and Andrew Rasiej, an owner of Irving Plaza, a large live-music club near Union Square, organized a "Quality of (Night) Life Forum" at Bottom Line in 1996 (Bumiller 1996). "Is there a campaign to drive out clubs to make the city safe for Disney?" Hershkovitz asked the participants in his opening remarks, referring to the city government's huge subsidy and incentives to the Disney corporation to locate in the refurbished Times Square. Answering in the affirmative, club owners present at the forum furiously vented about their struggles with residential communities and city authorities. Politicians, such as Borough President Ruth Messinger of Manhattan, and City Councilmember Kathryn Freed, who is a prominent anti-nightlife activist, were also present, and they firmly called on the owners to comply with laws and become "good neighbors."

However, being a good neighbor now involved a very substantial capital investment. Bars started to invest in private security measures, such as hiring bouncers—professionalized members of large private security companies—in order to control the crowds inside and outside their establishments (even in public streets), which was encouraged by the police and the municipality (Ocejo 2009: 10–11). In the case of Pravda in SoHo, where celebrities such as Brad Pitt and Hugh Grant were reported to hang out, the opening of the club was at first opposed by SoHo communities, which cost the club $130,000 in delays and attorney fees. The owner of the club even accepted most of the residents' demands and stipulations. The club went so far as to give money to a woman living above the club so that she could have her windows double-paned for soundproofing (Bumiller 1996). With these efforts, community relations improved, the club becoming a favorite spot even for Councilmember Freed. This episode re-inscribed the fact that moneyed club entrepreneurs were better positioned to survive in the ruthless "quality of life" era than cash-strapped ones. Clubs like Pravda did not come to the "Quality of (Night) Life Forum" mentioned above, saying that they were already on good terms with resident communities, and that they should not be conflated with other "trouble" nightspots (ibid.). Statements like this underline how fissures emerged amidst the nightlife community, as clubs positioned themselves as 'good clubs' in contrast to other clubs that were 'bad clubs,' [11] reinforcing the anti-nightlife governmentality encapsulated within the discourse of good vs. bad neighbors. Nevertheless, a trade and lobbying organization formed mostly of club owners, took shape at this forum, which became the New York Nightlife Association (NYNA), a successor to the former NYCA, that had been created in the late 1980s (see Chapter 5).

Deepening Conflicts and the Decline of Nightlife

In September 1996, the city government created the "Quality of Life Hotline" in order to respond efficiently to quality of life complaints over the phone. Identifying nightclubs as "a magnet for drug sales, underage drinking, loud music, and other conditions which create an atmosphere conducive to crime" (Owen 1997a), the municipality also created the Multi-Agency Nightclubs Task Force in 1997 (from now on, the Task Force). The crackdown on nightlife businesses increased in frequency and severity during Giuliani's second term starting in 1997, too (interview, NYNA representative, personal communication, January 14, 2005). New York Governor George Pataki also proposed in 1997 technical changes in the SLA law that would make it easier to close down troublesome businesses (Owen 1997b). This tightened law enforcement gradually wreaked havoc on the nightlife and subcultural scene in the city, which was simultaneously assailed by cut-throat rental markets. Nightlife businesses in the hippest addresses were most severely hit, but as gentrification reached wider areas in the city and real-estate prices remained at high levels city-wide during this period, the divergences between different neighborhoods were fading (Rothman 1999).

Reinforced CB powers continued to make it so that setting up a new club was exorbitantly expensive. At the same time, "the mood inside dance clubs . . . chilled" due to heightened surveillance by bouncers (Owen 1997a). Promoters of hip-hop nights now found it hard to find a venue that would host their events in Manhattan, and one anonymous Chelsea promoter said that the police told him "We don't want niggers in the neighborhood" (ibid.). When Tunnel re-started its hip-hop party every Sunday in late 1996, the police set up a roadblock in the vicinity, and stopped and searched people who looked like they were going there in order to search for drugs, guns and stolen items (Buckland 2002: 135–36). Queer spaces shared this blunt enforcement, too. Neighborhood opposition to Edelweiss, a renowned transvestite club in Hell's Kitchen, helped to get the place padlocked in March 1997, on the grounds that it attracted prostitution (ibid.: 132). Gay men cruising around the waterfront bars in Chelsea also experienced increased arrests for public lewdness (ibid.: 133). Media coverage of nightlife continued its sensationalist tone. Jack Newfield and Allen Salkin at the *New York Post* made pitched calls for the protection of kids from the seduction of drugs, a vituperative discursive vanguard in what amounted, overall, to a crusade against dance clubs (interview, Frank Owen, a club critic, personal communication, July 14, 2005).

Figure 6.2 Sign posted in front of a bar on Avenue A. Photograph by the author.

Andrew Rasiej, a co-founder of the NYNA opined that the mayor's uncompromising approach toward nightlife businesses, especially toward dance clubs, was "a direct result of the booming real-estate market in Manhattan, and the gentrification of previously commercial neighborhoods" (Owen 1997a). Though insisting that the situation was something that nightlife was not entirely responsible for, Rasiej believed that dialogue between residential communities and nightlife could assist in solving snowballing conflicts over nightlife. He tried to promote a Good Neighbor Policy among the NYNA members, and he himself gave out his beeper number to residents living around his club, so that they would be easily accessible to him at times of trouble. At the same time, he sought to lobby the City Hall and elected politicians to initiate more pro-nightlife policies, and commissioned a study on the economic impact of the nightlife industry in the city, to demonstrate more officially the substantial contributions of nightlife to the city's economy. There was also reflexivity on the part of club owners, who acknowledge that the municipal crackdown on clubland also emerged from violence that actually existed in the clubland. There was a rising sense that owners should demonstrate that they were making efforts to comply with the increasing regard for quality of life. Sound Factory's owner, before he opened his club at a new location in the Clinton neighborhood, sent club members a copy of the code of conduct that declared drug consumption and trafficking off-limits inside the club. Member clubs of the NYNA started to post blue-and-yellow posters urging patrons to "be respectful of neighbors" on their doors (Span 1998; also see Figure 6.2 for one variant of them).

The Task Force shut down 50–60 businesses between August 1996 and August 1998 (Rhode 1998). The watchdog resident organizations often played a crucial role in bringing to the police's attention suspicious activities that might be going on within clubs. The Save Avenue A Society helped the authorities to shut down three businesses between June 1997 and February 1998 (Span 1998). One member of the Society, who diligently patrolled Avenue A every weekend night in order to check whether clubs and bars were toeing the line, said that she was "fed up with the Young and Pierced . . . who come down here thinking this is the East Village theme park and they can act out" (ibid.). The owner of Spiral, an East Houston Street music club, which was paying $8,000 in lawyer's fees to deal with fines and summonses written by the Task Force for minor violations like illegal postering, took issue with the approach of the Save Avenue A Society. Clubs like his, he asserted, pioneered the gentrification of the drug- and crime-ridden Alphabet City. Hogs & Heifers, a honky-tonk bar in the East Village, was padlocked and fined for illegal patron dancing, and the owner paid nearly $150,000 in fines, legal fees, and eventually, renovations, in order to get the necessary permit for a cabaret license (ibid.). While law enforcement was to a certain degree indiscriminate to the types of establishments that they were writing citations for—whether they catered to gay, heterosexual, black, Latino, small, big, affluent, or poor contingents—it was usually well financed entrepreneurs who managed to survive in this anti-nightlife environment.[12]

By this time, New York bars and clubs had become familiar with the pattern of the Task Force's raids. The owner of Crossroads, an Upper East Side rock bar, makes plain their scale:

> It was late '97, early '98. They had about 40 officers come in a city-wide raid, representing the police department, the fire department, the health department, the vice squad, the consumer affairs department, and a legal team. You'd think they were making a Mafia bust . . . [They] make you shut the music off, turn the lights up. If you were a customer, would you stay? (Rothman 1999)

In this case, the Task Force was not able to find anything illegal in the club except an ice scoop illegally touching an ice cube, for which the Task Force wrote citations, and the charges for which were later dropped. The timing of these kinds of raids were of significance, as well as their scale, often coming during peak periods, on Friday and Saturday nights. Friday raids could be particularly expensive for club owners, as it often meant they were forced to close for the entire weekend, generating major financial losses (interview, NYNA representative, personal communication, January 14, 2005).

On the other hand, landlords, who aspired to have more profitable tenants, often used lease violations (like in the case of the Bitter End) or other violations as grounds for lease termination. The Chelsea club Tramps, for example, which offered a brilliant and eclectic mix of music, and which the police considered to be the least problematic club in the area, was closed in June, 1999 and the premises was replaced by an upscale restaurant-lounge. The lease termination originated from a minor lease violation that the landlord had previously ignored. The club owner suspected that the landlord wanted to cease the lease with the club, as the neighborhood now experienced gentrification, to acquire a tenant who would be able to pay higher rent (Rothman 1999).

In 1999, Coney Island High in the East Village—a small East Village rock venue, but with a reputation such that it was sometimes compared with the legendary CBGB—closed, as the owners could not make enough money to pay the high rents. The owners blamed the cabaret law enforcement for having caused financial costs to the club (Rothman 1999). In 2000, the NYNA representative had a meeting with city authorities to propose the "incidental dancing" exception to the cabaret law—that is, a provision that a simple incidental dance among just a few people in a business not licensed as a cabaret, such as an ordinary bar, should not be subjected to the cabaret law regulation. During the meeting, inspectors defined dancing as "moving rhythmically" or "gyrating up and down," and even took out a dictionary and pointed to various definitions of dancing, including "twitching" (Katz

2000). It was crystal clear that the municipality would not allow an exception to dance regulations.

The interpretation of the law according to such literal definitions of social dancing hit live music clubs hard, in particular. During punk music performances in clubs, for example, audiences may not necessarily dance to the music, but they may "twitch" or rhythmically "gyrate up and down" to the music; they, now, became subject to social dancing regulations. The history of the cabaret law, thus, proved ironic. As demonstrated in the previous chapter, the provision of the cabaret law that was discriminatory to specific types of live music venues was eliminated a decade earlier. At that time, in the *Chiasson* court, the Musicians' Union argued that social dancing establishments were different from live music venues, and implicitly moved to second the municipality's decision to tighten cabaret law regulations over dance venues, while removing regulations over live music venues. It now backfired to the disadvantage of live music venues, as the social dancing provision of the cabaret law now had a detrimental "domino effect" (Katz 2000) on live music clubs.

The indiscriminate manner by which officials came to attempt to apply the law was such that even instances of performance dance were faced with action by city officials. Officials fined, for example, the Swing 46 Club for having a swing dancer perform in front of an audience, although the charge was later dropped after lawyer Paul Chevigny petitioned to the DCA that performance dancing, unlike social dancing, was exempt from the cabaret law regulation (interview, Paul Chevigny, personal communication, September 16, 2004). The DCA, however, continued to press such charges, requiring owners, for example, to acquire a cabaret license if it wanted to have even a single belly dancer (Letter sent by an Assistant General Counsel at DCA to a bar/restaurant owner, 2003). These may be mishaps that are not uncommon in the bureaucratic process of law enforcement. But it may also show the logic of expediency that enforcement agencies have been working under, that is, the ways in which any available laws were adopted and enforced without first contemplating due process and procedural precision, in order to immediately respond to the complaints of residents and to demonstrate, at all costs, the effectiveness of agencies in clamping down on nightlife.

More frustrating was that the cabaret law application system often made it difficult for even a business located in as-of-rights zones to acquire a license. The story of Baktun on West 14th Street in Chelsea is well known in this regard. Baktun, a club with a 180-person capacity, opened in 1999 with a focus on cutting-edge electronic music and digital arts. It was, from the beginning, set up for dancing. When the Task Force first visited the club in 1999, the club did not hold a cabaret license, and was duly ticketed. As the club was located within an as-of-right zone for dancing, the owner, Phillip Rodriguez, applied for a cabaret license. The application sat in limbo somewhere in the Department of Buildings (DOB) for sixteen months. Rodriguez had already spent $70,000 upgrading the fire alarm and installing a sprinkler system, and further lost 50 to 60 percent of the club's gross income for eighteeen months straight as a net result (Romano 2002).

Throughout the process, the impression that Rodriguez got about DOB officials was that they "did not want to approve applications that included plans for dancing" (Brief, Philip Rodriguez, *John Festa et al. v. New York City Dept. of Consumer Affairs et al.*, 2). While Rodriguez was awaiting the DOB to get back to him with the approval, he received four more summonses from the DCA and NYPD for violations of social dancing provisions, which meant further fines. As a result, Rodriguez blocked any open spaces with furniture, and put a security guard in the middle of the dance floor in order to prevent patrons from coming to the dance floor to dance. "I don't produce anything creative anymore" (Katz 2000), Rodriquez complained, his hands tied by bureaucracy and expenses incurred for legal services. In early 2001, the DOB finally approved the floor plan and other applications made from Rodriguez for Baktun. Following this, the club had to go through an application for a Certificate of Occupancy (C of O), for which the club had to clear all the violations that had occurred before the club was opened on the spot, and also had to have the landlord clear all violations that had occurred in other parts of the building. All this completed, Rodriguez had to seek the DCA's approval, and had to wait to hear from the CB. Finally, he was granted the cabaret license. Shortly after, however, 9/11 occurred, and Baktun's business dropped. He had already lost a great deal of money during the process, and his lease was some three years from renewal. He finally sold the location to the management of Lotus, a nearby mega dance club (Brief, Philip Rodriguez, *Festa v New York City Department of Consumer Affairs*, 3).

The shift of the ownership signified the gradually emerging, but by then palpable reality, that the cabaret law regime, if not purposefully, had nurtured uneven development within the city's club industry.

> Baktun is directly across the street from Lotus, but the divide couldn't be greater. There are no red velvet ropes in front of Baktun, no celebrities clamoring to get in, only clubbers seeking a dose of good music made more palatable by the club's low cover charge and relatively cheap drinks. "We get along together," [Rodriguez] says of [the owner of Lotus]. "But it's a different industry." (Romano 2002)

The owner of Lotus was then the president of the NYNA, which represented affluent mega clubs like Lotus. Discontents over uneven development in clubland was simmering, and would soon erupt into conflicts between different sectors of club industry—issues I cover in the next chapter. Baktun was joined in its closure by Wetlands, which shuttered its doors in September 2001 to make a way for condo developments after thirteen years of live music performance and environmental activism. Underground club actors were in shock, and on-line music webzine *World Beat* commented in its eulogy to the club that the loss of clubs like Wetlands would add to the predominance of "trendy lounges and carpeted live music venues" in the city's clubland.

CONCLUSION

Nineteen nineties nightlife regulation inherited the regulatory regime of licensing and zoning from the 1980s, but a newer set of "legal complexes" also emerged. New laws both at the municipal and state level were proposed, and some passed, that resulted in the further tightening of licensing and zoning requirements of not only dance clubs, but other troublesome nightlife businesses, too. One of the prominent breaks between the 1980s and 1990s was that residential communities were increasingly well organized, and their concerned voices allowed to inform the licensing process of nightlife businesses in their neighborhoods. It was clear that elected politicians gravitated toward supporting residential communities' anti-nightlife initiatives, while summarily ignoring concerned voices about the waning nightlife scene that had enriched alternative subcultures, in particular. The Happy Land fire reinforced the rationality of the knowledge-claim that safety in nightlife businesses featuring entertainment, especially those offering social dancing, is compromised due to the "carnivalesque" atmosphere of these clubs induced by the combination of alcohol, music, dance and drugs. This rationality still appears in various techno-bureaucratic venues, justifying the authorities' stringent enforcement of nightlife regulations in response to even very minor violations, even in cases where relevant problems have had little to do with safety compliance.

As was seen in the contrasting discourses of "good, responsible vs. bad, irresponsible" neighbors, nightlife problems were framed in techno-bureaucratic language as originating from the mismanagement of nightlife business owners of their premises and incompetence in controlling their patrons, and not at all due to the gentrification of formerly derelict neighborhoods. Trapped by the bureaucratic language, nightlife businesses had to play towards being 'good' and 'responsible' businesses. But this turned out to be highly expensive, because businesses had to spend large sums employing more security personnel, purchasing sound-curbing equipment, providing indoor waiting spaces and so forth. It turned out that moneyed entrepreneurs were better able to survive in this milieu, and small and underfinanced businesses—often those that tried alternative, experimental performance, music and dance—started to disappear from the city's nightlife scene. Faced with these material realities, many business owners sought to attract less controversial patrons and more affluent, upscale clientele.

Noisy and disorderly nightlife businesses had become so stigmatized that anti-cabaret law activists, as will be shown in the next chapter, found it hard to defend noisy and disorderly businesses. What this whole process bespeaks is that the nature of nightlife had changed. What *kind* of nightlife would it be if people are expected to be quiet and are under constant surveillance? If clubs have to constantly ask bands to reduce the sound of the music they play, keep patrons from rhythmically moving to the sound of the throbbing music, and stop patrons from socializing outside the club, what *kind* of subculture will take shape (if they take shape at all) in these nightclubs? Rich subcultures that have defined New York were born in clubs characterized by grime, noise, sleaziness and zany wildness. The changing geography of nightlife in the city may portend that this will no more.

The crackdown on nightlife also involved a grave infringement on people's rights to dance together. Through its stringent enforcement of the cabaret law based on nitpicky interpretations of its social dancing provision, the Giuliani administration demonstrated how a government urgently pursuing high quality of life for new gentries could execute an outright dismissal of civil liberties. The municipal government appropriated popular revanchism and established policies, laws and campaigns through which the blatant jettisoning of civil liberties and crackdowns on the disenfranchised could be legitimized. These policies did not go unchallenged, however. The growing sense that good, organic nightlife had been destroyed in the city due to cabaret law enforcements also pushed people in the city's subcultural scene to come out and voice their demands for changes in the cabaret law. It is to the activism of these groups that we now turn in the next chapter.

7 Voices for Change

From 2002 Onwards

In response to governmental crackdowns on venues carried out on the basis of the three people dance provision, clubs and bars developed tactics to evade cabaret law violations (Healy 2003; Steinhauer 2004). Illegal cabarets like Cooler in the meatpacking district (which is currently defunct), or Pyramid in East Village (which is no longer illegal), used tactics prevalent among "stealth" gay clubs in the 1960s. Once the presence of the so-called MARCH (Multi-Agency Response to Community Hotspots)[1] was detected in the neighborhood, a message would be hurriedly conveyed to the DJ of the bar. Once the MARCH squad stepped inside the venue, they would find people chatting with each other over drinks while lounge music played in the background, with no dancing in evidence.

If such De Certeau-ian types of "tactics" to subtly transgress and subvert the established rules marked one kind of strategy, a more aggressive kind of "high politics" evolved at the end of the Giuliani administration through to the early Bloomberg administration in order to challenge the cabaret law, in the form of Legalize Dancing in NYC (LDNYC). LDNYC sought to galvanize popular support for their causes with a First Amendment-oriented mantra, i.e. one stating that social dancing consisted of expression that ought to be protected. The political milieu in NYC had become somewhat propitious to these anti-cabaret law activisms, as the Bloomberg administration seemed to understand that the three people dance provision not only damaged the administration's regulatory legitimacy, but also was not effective in combating the "real" problem of nightlife—noise. LDNYC, and the still-operative New York Nightlife Association (NYNA, that was introduced in the previous chapter), contended that the municipal crackdown on the city's nightlife was responsible for its recent downturn. But the NYNA, contrary to LDNYC, opposed the abolishment of the three people dance provision of the cabaret law. This divide between these two groups over the issue of the cabaret law mirrored, on the one hand, the uneven development between the different segments of the club industry that has become salient in recent years and, on the other, the transformation of the city's club culture. In this chapter I show that the opposing approaches that each of these organizations took to some extent reflect different interests coming especially from different representative classes of the membership of each organization. In particular, I show that the NYNA may represent the interests of affluent business owners in particular, and has been party to the gentrification of nightlife in the city, an ironic position given that the NYNA has also been contesting the crackdown on nightlife by the city's gentrification regime.

This chapter elaborates on how dissimilarities between these two groups, LDNYC and the NYNA, played out in their political struggle with the municipality. As part of this discussion, I highlight the multiple factors which conditioned the courses and outcomes of political struggles waged by these two organizations. The new nightlife regulations proposed by the DCA in 2004 and controversies that erupted in their wake showed how these multiple factors were at work in shaping political outcomes of pro-nightlife activisms. In particular, I attend to the undemocratic structure of the licensing and zoning system of the municipal entertainment regulations, the popular stigmatization and devaluation of the cultural importance of "noisy" nightlife businesses, and polarization between different sectors of the nightlife industry. The evidence marshaled demonstrates how difficult it is for politically marginalized and socially stigmatized groups of people to challenge zoning as well as licensing regulations.

In addition, I show how LDNYC's single-issue focused activism backfired in the course of the struggle. LDNYC zeroed in on anti-cabaret law campaigns, which advanced an important civil liberties issue, for sure. However, this also represents a limited challenge to the anti-nightlife regime and, more fundamentally, to the market discipline exerted over nightlife businesses through gentrification. This is because LDNYC's activism has been all about protecting civil liberty from undue governmental attack, but has not extended to fighting the gentrification *per se* that has routinely engendered crises of nightlife since the early 1980s, led the municipality to abuse the social dancing provision of the cabaret law to suppress nightlife, and has subsequently widened the disparity between unevenly financed nightlife businesses. Based on this examination, I argue that what is needed for pro-nightlife formations is, in addition to fighting against governmental suppression of nightlife, a more robust political response that involves fighting gentrification *per se*, as it is this that has posed a fundamental threat to a vibrant and culturally rich nightlife. Both NYNA's and LDNYC's activism has shied away from this more robust and comprehensive politics, which I argue should be the foundation to usher in a more egalitarian and democratic nightlife in the city.

BLOOMBERG: NOISE FOCUSED

With Michael Bloomberg elected as mayor in 2002, a hope emerged in clubland that nightlife regulation in his administration would not be as excessive as that practiced during the Giuliani administration.[2] Indeed, during the first few months of the Bloomberg administration, inspections by the Multi Agency Response to Community Hotspots (MARCH) seemed to drop off. Beginning in the spring of 2002, however, MARCH's activities increased dramatically. Between then, through to November 2002, enforcement actions by the MARCH increased 35 percent over the same period in the preceding year (Steinhauer 2002). Other statistics reported that between April 2002 and June 2003 MARCH issued nearly three thousand tickets for unspecified violations by nightlife venues (Snyder 2005: 255). The hiatus in nightlife business inspection between late 2001 and early 2002 may have been a mere aftereffect of 9/11.

Among the above violations were ones written out to venues that illegally allowed more than three patrons to dance in their premises. In June 2002, after the MARCH squad found a few patrons dancing in the Slipper Room in the Lower East Side, the club was padlocked for a second time and fined $30,000, although this was later reduced to $6,000 (Romano 2002). Inspectors padlocking the club told the owners that one more violation would mean closure of the club. Civil rights lawyer (and former director of the New York Civil Liberties Union) Norman Siegel took up the club's case. In a meeting with DCA officials, the owners of the club represented by Siegel were told that the DCA did not want another Happy Land-type fire to occur, and that this was the reason for the stringent punishments imposed on the Slipper Room (ibid.).

Despite the same old patterns and rationales of cabaret law enforcement, there was a difference in the approach that the Bloomberg administration took in the matter of the cabaret regulations from that of the Giuliani administration. The Bloomberg administration seemed to understand that using the three people dance provision as a means to control primary problems of nightlife and secure the quality of life of residential neighborhoods was seriously flawed. For example, complaints by residents regarding noise generated by nightlife were soaring, putting pressure on the municipality to scale up the enforcement of laws available to curb the noise problem. The effect, however, of enforcing the cabaret law in solving the problem was found to be negligible, and even counter-productive as the popular discontent at the limitations on being able to dance started to threaten to undermine the legitimacy of the city's regulatory regime. In this context, there were rumors that the mayor's office, in collaboration with the DCA, was considering reforming the cabaret law (Huden 2002).

At the same time, the Bloomberg administration focused on fighting the city's noise pollution. According to the Mayor's Office, among the average of ninety-seven thousand complaints that came through to the Quality of Life Hotline (which would be changed to the 311 line in 2003) each year, approximately 85 percent were about noise. The commissioner of the Department of Environmental Protection (DEP) said, "noise complaints in the city are increasingly an indicator of a lack of civility and urban disorder"

(Office of the Mayor 2002). The mayor's office also maintained that getting tough on noise pollution would not only reduce residents' sleepless nights, but also have a crime reduction effect—the same effect that, Bloomberg argued, the former Giuliani administration's efforts to eradicate squeegee men, turnstile jumpers, and other petty offenders had produced (Burger 2002). Basing these new policies on the same assumptions of the "Broken Windows" thesis (cf. Wilson and Kelling 1982) that animated the Giuliani administration's Quality of Life campaigns, Bloomberg's combat against noise assumed a causal relation between noise and crime rates that was questionable at best.

The City developed and implemented initiatives to regulate noises, such as Operation Silent Nights, and realized that the Noise Code of the city was flawed in a number of respects. Most serious of all was the subjective nature of the police's judgment in understanding the level of "excessive" noise and the writing of summonses according to these judgments. This subjective nature was the source of endless disputes between the police and nightclubs, and the main reason that courts had annulled a number of violations that the police had brought to the court. This shortcoming pushed the municipality to seriously deliberate on overhauling the thirty-year-old Noise Code, and to establish a more objective, comprehensive, and up-to-date set of regulations. This move, along with the administration's rising awareness of the ineffectiveness of the cabaret law in clamping down on the noise problem, created a favorable opportunity for anti-cabaret law activists. To be sure, the Bloomberg administration was willing to hear the voices that opposed the cabaret law.

EMERGENCE OF ANTI-CABARET LAW ACTIVISM

The three people dance provision of the cabaret law during the Giuliani administration spurred protests from dancers, promoters, club owners, and other club industry types. In the late 1990s, these protests coalesced into more organized activism. The Dance Liberation Front (DLF), composed of dancers and performers, was formed in 1998, and hosted creative performances in the city's public spaces, delivering messages to the public about the absurdity of the cabaret law regulation of dance, and about dance as an inalienable right that ought to be protected from this type of regulation.

While the DLF's activism was more focused on creative performative protests, the other major anti-cabaret law organization, Legalize Dancing in NYC (LDNYC), formed in 2001, combined performative protests with popular campaigns. LDNYC originally grew out of a music collective called Mishpucha, a music industry community of 120 record label people, artists, DJs, promoters, managers and writers who championed innovative music and music activism (interview, LDNYC leader A, July 27, 2002). In August 2001, Mishpucha organized a forum to discuss the problems of the cabaret law (Neuberg 2001). Two hundred people congregated at this forum: business owners harassed by the authorities for allowing patrons to dance, club patrons, members of the DLF, activists against the Reducing Americans' Vulnerability to Ecstasy Act (a.k.a. the RAVE Act that was, many suspected, designed by the US Congress to crack down on outdoor rave parties), lawyers that represented nightlife establishments, members of the New York City Late-Night Coalition (NYCLNC, a group of people in the club industry formed to raise voter awareness regarding problems with the City's crackdown on nightlife during the mayoral election campaign in 2001). Later, leaders of the NYCLNC, Mishpucha, Action New York (another pro-nightlife organization) and the DLF created an alliance called Dance the Vote (*www.undergroundarchive.com*). The alliance contacted mayoral candidates to influence their nightlife policies, especially those concerned with the cabaret law. In September 2001, the alliance also helped bring about a victory in the primaries for then-City Council candidate Alan Gerson, who was running in Council District 1, which covers 90 percent of lower Manhattan.

By the time the election campaign wrapped up, Mishpucha decided to mobilize interested people and create a pro-nightlife organization. While there were multiple causes of nightlife conflicts, Mishpucha decided to primarily engage in fighting to reform regulations related to social dancing under the cabaret law. Soon, Mishpucha formed an anti-cabaret law group called No Dancing Allowed, and was later renamed Legalize Dancing in NYC (LDNYC). LDNYC consisted of cultural critics, music-industry types, writers, performers, promoters, and business owners. LDNYC organized a few public events, including the No Dancing Allowed Walk in the Coney Island Mermaid Parade, and for a while it ran a resourceful (but currently defunct) website that explained the provisions and the terms of the cabaret law (*www.legalizedancingnyc.org*). LDNYC started to work with Councilmember Alan Gerson more intensively after his election to propose the reform of the cabaret law in the City Council. This effort was later joined by two New York Civil Liberties Union lawyers—Paul Chevigny, who had represented the Musicians' Union's when it challenged the cabaret law in court in the late 1980s (see Chapter 4), and Norman Siegel, who was at that time representing the Slipper Room. The members of LDNYC, Alan Gerson, Paul Chevigny, and Norman Siegel actively contacted City officials, especially officials at the DCA, to negotiate a political settlement that would solve the problem of the cabaret law. They also extensively mobilized local media to make it more publicly known that social dancing was being unduly regulated in the city.

The recognition that social dancing is an important right was the leitmotif of LDNYC activism. The right to social dancing, this group argued, is equal to a right to sovereignty over an individual's body, a right to collective celebration of the body, and a right to a particular type of socialization. As the cabaret law enforcement was especially inimical to small, alternative clubs, the rights claim to social dancing was also a claim of people's right to access to a space where diverse alternative subcultures were experimented with, performed and thrived. Cracking down on this access would seriously stifle the cultural verve of the city (www.legalizedancingnyc.com, last accessed in July, 2004). The cultural and social cost caused by the enforcement of the cabaret law was too high, LDNYC argued. LDNYC further argued that in curbing the nuisance effects of cabarets, such as noise, crowding, safety issues, regulating social dancing through the enforcement of the cabaret law would not do any good. Social dancing had nothing to do with these nuisances, LDNYC contended, and these nuisance effects would be more effectively curbed through the enforcement of noise, parking, and safety ordinances.[3] Despite the fact that social dancing was not connected to such nuisance effects, LDNYC argued, the administration continued with its regulatory practices because it was the easiest way to regulate raucous nightspots. Social dancing, LDNYC argued, was not a trivial activity that could be randomly utilized by an administration for such conveniences. Dancing is fundamental to human life and culture, and, therefore, an important right (interview, LDNYC leader A, July 27, 2002).[4]

This type of rights talk resonated with the public, and media coverage was favorable towards LDNYC's activism. However, the LDNYC's focus on the cabaret law also caused occasional internal disputes among its members. LDNYC leaders wanted to develop the group as a one-issue interest group (email correspondence, LDNYC leader B, February 21, 2007; various communications in the LDNYC listserve), and tried to prevent other issues from being included in the LDNYC agenda. The LDNYC leaders, in particular, thought that including other issues—even if they equally harassed the nightlife scene in the city—might dilute LDNYC's sole focus on liberating dancing. Under this logic, for example, discussions of the possible impact of the upcoming smoking ban in restaurants and bars, which a number of LDNYC members were concerned about given the potential to increase noise at street level as patrons would stream outside to light up, thus harming relations between nightlife businesses and neighboring residents (e.g. LDNYC listserve, August 15, 2002),[5] were rejected as irrelevant and eventually banned from discussion on the listserve. In the same way, LDNYC members were discouraged from developing discussions on what LDNYC should do about the municipality's overall offensive on noisy and "disorderly" nightlife businesses if they were not related directly to cabaret law enforcement. Reports on such crackdowns occasionally appeared on the LDNYC listserve, but this was rarely developed as a coherent part of LDNYC activist agenda. LDNYC leaders continuously reminded the members that reforming the cabaret law should be the focus of LDNYC activism, and the problems related to the regulations of the nightlife premised upon safety, noise, etc. should be of secondary concern.

By mid 2002, the matter of the singular focus of LDNYC's activism came to a head. Councilmember Alan Gerson discussed the issue with both DCA

Commissioner Gretchen Dykstra and Mayor Bloomberg's "inner circle" of advisors, who sympathized with the problems of the cabaret law that were raised by LDNYC activists, but who were still reluctant to get rid of social dancing regulations at the risk of defiling quality of life. Gerson sought to persuade LDNYC leaders that the best political route to take in order to reform the cabaret law would be to simultaneously support the reform of the city's Noise Code. Gerson argued that legalizing dancing would raise concerns about possibilities of increased noise both inside (loud music) and outside (rowdy patrons) venues, and therefore, deregulating dancing would stand a chance only if coupled with noise regulation reform (ibid., August 23, 2002). There was internal opposition among LDNYC members regarding Gerson's proposal, as it sounded as if LDNYC would be endorsing more harassment of nightlife based on new stricter noise regulations. One member lamented that the proposal would relegate the legal dancing movement to being a "footnote to greater 'Quality of Life' enforcement" (ibid., November 10, 2002). LDNYC leader C also said, in an interview with *Villagers* (Anderson 2002), that some LDNYC members were reluctant to couple repeal of social dancing regulations with legislation of stricter noise codes, because that would imply that LDNYC consented to the idea that deregulating dancing would bring more noise to neighborhoods. In the end, LDNYC decided to push for the removal of dance regulations without accepting any conditional tightening of the noise code. LDNYC maintained that they ought to respect the *existing* quality of life laws that regulate noise and safety, but the existing cabaret law that criminalized dancing should be abandoned, and that eliminating the discrimination against social dancing should not be coupled with the tightening of other nightlife regulations, such as noise rules (DCA 2004: 75, 209, 233).[6]

Even though LDNYC dropped the Gerson plan, the fact still remained that LDNYC's activism was limited to liberating dancing, while divesting its agenda from other issues that hounded nightlife venues. LDNYC was able to gain tremendous traction with claims regarding the government's abuse of fundamental rights associated with dancing, but such framing also limited the potential of LDNYC's activism because it could not develop a more comprehensive challenge to the city's loss of a noisy, boisterous and "carnivalesque" nightlife that were invaluable to cultural creativity. As well, LDNYC's endorsement of (existing) Quality of Life policing may have resulted in endorsing the legitimacy of the governmental criminalization of noisy and disorderly venues. This was also related to the lack of a discussion and debate among LDNYC members over the more fundamental force, i.e. expanding gentrification, which had put the city's nightlife in constant conflict with residential neighbors and the city government, and pressured the municipality to use any available laws, including the cabaret law, to combat nightlife. In LDNYC's official language, the enemy of the city's nightlife and subcultural scene was the cabaret law that trampled upon basic freedoms, as well as the municipality that enforced the law, but not the gentrification of urban space. LDNYC hardly reckoned with the ironic prospect that subcultural places that it was trying to protect by reforming the cabaret law would continue to suffer even after the elimination of the social dancing regulation *due to* hiked rents or the crackdown on noises of nightlife by the municipality which would be ostensibly enforced to enhance quality of life in gentrifying neighborhoods.

However, challenging the entire gentrification project would have been too ambitious a project and may have cost LDNYC the loss of popular support. The members may have had to strategically choose focusing on the cabaret law as an effective means of political mobilization and popular support. A key LDNYC member said:

> I suppose it seemed like the cab[aret] law was the most obvious flashpoint in the whole mess. Pretty hard to campaign in favor of loud noise, but when you frame the argument in terms of justice and rights and the [government's] abuse of the law, then you can gain some traction. (email correspondence, an LDNYC member, February 22, 2007)

That is, incorporating into LDNYC's activism agenda the other issues that troubled the city's nightlife, such as noise conflicts, may have not elicited as wide a popular support for its activism as the focus on curbs on dancing as violations of First Amendment rights could. Noisy and disorderly nightclubs had been stigmatized as a menace and even as a crime, and hardly thought of as important cultural institutions in the civil society, and therefore, acting with the explicit objectives of saving them would have evoked less popular support.

One of the LDNYC leaders also said:

> LDNYC had no broader plan for an activism to challenge the overall governmental regulations on the subcultural scene and the city's nightlife . . . [U]ltimately we were a one-issue interest group. We wanted the city to let people dance wherever they wanted. That was all. (email correspondence, LDNYC leader B, February 21, 2007)

That is, the single focus of LDNYC could be ascribed to the fact that LDNYC was set up by a gamut of interested citizens as well as music/dance creatives who were enraged by the egregiousness of the cabaret law regulation, and therefore merely wanted to fight against the law, as opposed to fighting over the broader issue of aggressive nightlife policing. And LDNYC's financial and political power was too small to tackle such a broad issue. In retrospect, even reforming the cabaret law turned out to be a tough task that eventually failed (as I describe later in this chapter). It would thus seem clear that it would have been beyond the capacity of LDNYC to work on fighting broader issues related to gentrification and the Quality of Life policing that beleaguered the city's nightlife (email correspondence, LDNYC leader B,

February 21, 2007). Whatever the reasons for them, the limits of LDNYC activism that I have discussed here rose as a thorny issue in the chaos following the new nightlife proposal written by the DCA in 2004 to replace the cabaret law.

THE PRO-NIGHTLIFE ACTIVISM OF THE NYNA

Since 1996, when it was founded, the NYNA has been described as an organization that represents nightlife business in the city. What made the NYNA controversial among nightlife actors was that the group did not participate in the groundswell of anti-cabaret law activism that emerged in the late 1990s; as a matter of fact, the NYNA was officially opposed to abolishing the social dancing provision of the cabaret law. Instead of finding fault with the cabaret law regulation, the NYNA was critical of clubs, bars and lounges that were operating at the borderline of legality. The president of the NYNA publicly criticized clubs that illegally allowed patrons to dance, and even went on to name the names of illegal clubs—mostly competitors of his club, Lotus—in one instance, in an interview with Page Six in the *New York Post* (Johnson 2002). In another interview, the president justified his critical stance towards owners of certain clubs, saying that "[b]ecause [owners] have lied about what they are going to do, [community boards] are going to assume you are lying also" (Romano 2002, original brackets), ultimately jeopardizing the relationship between clubs and Community Boards (CBs).

While the NYNA was willing to support, even officially, proposed exceptions for incidental dancing, it objected to amending the cabaret law in the way proposed by LDNYC.[7] Some NYNA members argued that the repeal of the three people dance provision of the cabaret law was not fair to those entrepreneurs who had already invested enormous amounts of money to acquire the cabaret license (interview, NYNA representative, January 14, 2005). While such an argument might sound reasonable, according to Tricia Romano at *Village Voice* (2002), others pointed out that the cabaret license regime had created a "club cartel" composed mainly of club owners with abundant financial power (most of whom were members of the NYNA)—the only entrepreneurs who could open dance clubs in an environment marked by high real-estate expenses, extreme Quality of Life policing, rigorous anti-nightlife activism on the part of residents and onerous expenditures involved in acquiring the cabaret licenses. The clubs belonging to this cartel could effectively protect their own businesses, and even maintain a monopoly by reinforcing the present cabaret law regime that would practically make it more difficult for less affluent entrepreneurs to enter into the competition. The cabaret law regime, thus, was partly responsible for the widening unevenness within the city's clubland. It was further reported that many of expensive bars and mega clubs had been more concerned with providing so-called "bottle service" [8] to celebrities, tourists, B&T youngsters,[9] or newly transplanted dotcom corporation workers, and routinely displayed little concern for either music or dance. For these affluent clubs, the cabaret law and its possible violation of the principles of the First Amendment, together with the law's vicious impacts on the development of subcultures, was something to shrug off because this was not their primary concern. Romano, in a recent communication with the board of the NYNA, has observed:

> Looking at all the conversations and e-mail exchanges with [the president and the representative of NYNA], there are many mentions of "investment," "business," "zoning," and "safety." There are a few mentions of dancing or the First Amendment and freedom of expression. But not once do they mention music. (Romano 2002)

The other reason why the NYNA was reluctant about agitating for the abolition of the cabaret law was because of the problem of zoning involved in the cabaret licensing process. It seems that the NYNA understood that in order to ease the three people dance restriction, the City Planning Commission (CPC) would have to reform its zoning regulations (and the associated building codes) of the cabaret law, which would not involve the DCA, because the latter was only in charge of issuing the cabaret licenses (in contrast, LDNYC had mostly been working with the DCA).[10] The NYNA argued that having the CPC modify zoning regulations would be a long, convoluted, unpredictable and eventually losing battle. As residential communities considered the zoning layout to be critical in maintaining the quality of life of neighborhoods (interview, NYNA representative, personal communication, January 14, 2005), an attempt to ease the zoning regulation would amount to opening a can of worms. In particular, communities would think that deregulating dancing in the zoning resolution would bring about a completely different land-use and different types of populations to their neighborhoods (even though this amendment would be a modest one), which would no doubt spark negative reaction by residents (email correspondence, NYNA representative, May 13, 2007). Politicians, therefore, would not risk jeopardizing their electoral support in order to accomplish so humble a goal as allowing people to dance. In that vein, touching the zoning issue would put NYNA nightclubs at risk of spoiling their relationship with communities that they had painstakingly built up over the years (statement by NYNA representative in Anderson 2002). Equally, nightlife businesses in general were not considered as a legitimate constituent of the "public interest," legitimate enough to move the CPC to consider changing the zoning regulations for the betterment of nightlife businesses.

The above were the reasons why the NYNA focused on issues unrelated to dance as (sub)cultural expression or as a First Amendment issue. Instead, they focused on issues such as pressuring the NYPD to grant approval for the use of Paid Details Units (PDUs) by dance clubs and bars.[11] The hiring of paid detail officers, the NYNA asserted, would be highly effective in curbing street disorder caused by nightclub patrons, because a uniformed NYPD officer outside an establishment can more effectively wield the authority to make people lower the level of noise they make on the street than, say, private security guards employed by individual businesses (cf. comments by the president of NYNA in DCA 2004: 81–82).[12] However, the police chief had been rejecting the use of PDUs by nightlife businesses due to reasons of liability and conflicts of interest. The other effort exerted by the NYNA was to have the city government ease or remove the smoking ban. The NYNA contended that this ban not only caused businesses to lose patrons to its New Jersey counterparts, but also increased noise levels on the street due to the smokers who come out to the street in order to smoke (Figure 7.1). In its rally against the smoking ban, NYNA made an alliance with Taverners United For Fairness (TUFF), a coalition of bar and restaurant owners to protest mayor Bloomberg's smoking ban. In an effort to demonstrate the economic resourcefulness of the city's nightlife industry, in 2002, NYNA commissioned a visitor profile and an economic impact

Figure 7.1 NYNA President speaking at the "Can the [Smoking] Ban" rally outside City Hall. Source: NYNA Newsletter, Fall 2003.

analysis of the nightlife industry, which was published in January 2004. According to the report (NYNA 2004), the nightlife industry "generated an estimated $9.7 billion in economic activity, $2.6 billion in earnings (primarily wages) and 9,550 jobs" in the city, and "the industry contributed an estimated $391 million in tax revenues to New York City and an additional $321 million to New York State."

There were NYNA members that did not agree with the NYNA's official objection to the abolition of social dancing regulation of the cabaret law. According to an NYNA member who was also a member of LDNYC (email correspondence, February 14, 2007), the NYNA board did not represent the voices of all the members in the matter of the cabaret law, and the board's decision on the cabaret law was swayed by "a handful of vocal, active members, usually the bigger name players who have more money, more at stake, and larger companies that afford them the time to spend on these things." NYNA membership fees were not high ($200 annual fee), but the decision-making process, it seems, was not democratic. When the board's interests conflict with those of the rest of the NYNA membership, it was the case that the NYNA's final decision would gravitate towards those of the board members (interview, New York-based DJ, personal communication, Dec 22, 2009). Not only were NYNA policies questionable in terms of its internal democratic representation, but how representative the NYNA has been of the city's club industry overall is questionable. It has never been clear how many businesses and what types of businesses it represents. The NYNA representative has always declined to inform me of this information. He told me only that the NYNA represents "enough" businesses (interview, December 11, 2009).

NEW NIGHTLIFE LICENSE

The DCA held a public hearing about the reform of the cabaret law in June, 2003. A massive number of people gathered for the public hearing, including experts in security and sound-proofing technologies, insurance companies, members of the NYC chapter of the American Planning Association, LDNYC, DLF (Figure 7.2), the NYNA, ordinary bar and club owners, lawyers that represented bars and clubs, members of the NYC restaurant associations, associations for tavern, hotel, and other relevant businesses, and members of a number of CBs. Discussions revolved around the problems of cabaret law enforcement, and recommended solutions towards nightlife conflicts. NYNA board members on the panel focused on enumerating the damages that had resulted from the smoking ban, the necessity of PDUs, the multiple efforts on the part of nightlife businesses to reduce trouble both inside and outside venues and also their willingness to work with residential communities, and the enormously positive impact of nighttime businesses on the city's economy. The NYNA board members did not address one word towards how to reshuffle the terms of the cabaret law so that social dancing would not be unduly suppressed. This contrasted with the argument advanced by Norman Siegel, another panelist on the same panel. Siegel discussed the constitutional entitlement implicated in social dancing, and argued that the government could not show a legitimate governmental interest in regulating social dancing (DCA 2004).

Right after the public hearing, the DCA commissioner, Gretchen Dykstra, and Mayor Bloomberg pronounced that dancing should not be regulated in the city, and stressed the need for a new regulation that aimed to combat noise and traffic congestion that was oriented towards nightlife spots rather than social dancing itself. In November 2003, the commissioner Dykstra announced that the DCA would not amount to a dancing police any more (Romano 2003a), and revealed the new proposal for the nightlife license. The new proposal included the following rules (DCA 2003):[13]

> Any public or private establishment located wholly or partially within a residential (R), C-1, C-2, C-4 or C-6 (1–4) zoning district with a capacity of 75 or more people which is opened after 1 a.m. that wishes to have continuous live or reproduced sound at 90 dBC leq. or higher must obtain a two year nightlife license.

Figure 7.2 Members of the Dance Liberation Front (DLF) offering a hokey-pokey performance during a break at the DCA public hearing in June 2003. Source: O'Brien, 2003.

> Any public or private establishment located wholly within a commercial (other than C1 or C2), manufacturing or special mixed use zoning district with a capacity of 200 or more people, which is opened after 1 a.m. that wishes to have continuous live or reproduced sound at 90 dBc leq. or higher must obtain a two year nightlife license.
>
> Every establishment that has a capacity of more than 499 must have one state-certified security guard for every 50 allowable occupants (plus supervisor) and deploy sufficient guards to maintain order outside the establishment when needed.
>
> If an establishment violates DCA license laws (including exceeding the self-certified maximum sound level) three times within two years, DCA shall have the power to padlock the establishment for between one and ten days.
>
> DCA may revoke a nightlife license only if the location is indicted for two of the following on different days within two years. These will include incidents of: homicide, assault, rape or attempted rape, possession of weapons, unlicensed sale of liquor, sale of liquor to minors, overcapacity, disabled sprinkler systems, exit signs or emergency lighting, blocked or locked exits, and two DCA padlocks.

Under the new proposal, then, dancing would not be criminalized in bars, restaurants and clubs, but instead, nightlife establishments would be regulated according to their noise level, crowds, and other criminal activities. In addition, nightlife businesses with a capacity of under seventy-five in residential zones (including light commercial zones), and ones with a capacity of under two hundred in commercial or mixed use zones would be exempted from having to acquire a nightlife license.

It became clear, however, that the proposed nightlife license law would impose even more taxing requirements on the city's nightlife businesses than the existing cabaret law itself. First, the eradication of the cabaret licensing process on the DCA's side did not mean that the most central part of (the problems of) the cabaret law—i.e. the zoning regulation of the cabaret law under the supervision of the City Planning Commission (CPC)—would go away. This meant that nightlife businesses located in places not zoned for cabarets would still not be able to have dancing under the new proposal. At the same time they would have to get a new nightlife license if they were over a certain capacity, and sought to stay past 1 a.m. It was obvious that inspectors from the Department of Buildings (DOB) would be the next dance police, inspecting violations of the cabaret law,[14] whereas the DCA would implement the new nightlife license law and inspect businesses according to that law (Romano 2003b). In response to critiques raised about this point, commissioner Dykstra said, "The DCA had a law it's supposed to enforce that doesn't address a problem, and that'll be true for the buildings department, too. There are folks who would like us to go farther outside our purview and that's not my job . . . We are not the Department of City Planning" (ibid.).[15]

In addition to this zoning quandary, it was argued that provisions of the new license law were too onerous for nightlife businesses. The NYNA was the major critic of the new proposal, citing its negative impacts. Three minor violations in two years, including something like a sidewalk left unswept a half hour after closing time, would allow the city to shut down an establishment for up to ten days. Further, two infractions in the "serious" category within two years would permanently revoke the license for a club. "If a place has two fistfights in two years and there are indictments—not convictions—stemming from that, your license can be revoked permanently," said the president of the NYNA (Tantum 2004). In addition, despite the DCA's declaration that the new nightlife regulation would only target the three problems of "The Big, the Loud and the Late" (DCA 2003), the NYNA argued that the regulation would in actuality involve most of the nightlife venues in the city. Contrary to the DCA's contention that most nightlife businesses had internal noise below 90 decibels based on its preliminary survey of fifteen venues in the city, the NYNA and TUFF argued that hundreds of nightlife businesses that were not subject to the cabaret law had interior noise levels of over ninety decibels, and therefore, they would all need to apply for the nightlife license (TUFF listserve, February 13, 2004). This meant that these businesses would need to invest in a sound engineer to measure their average decibel level, and also install costly soundproofing devices in their venues, which would induce them voluntarily to close their businesses at 1 a.m. The NYNA representative said, "There is no question this is a de facto closing of bars and clubs at 1 a.m [. . .]because no one can withstand the provisions of this law" (Steinhauer 2004).

The NYNA additionally accused the DCA of using the seemingly happy repeal of the cabaret licenses in order to subtly smuggle in even more draconian nightlife regulations. For them, the DCA was using this new proposal as a "Trojan Horse" (TUFF listserve, February 20, 2004), through which it could enhance its image by giving the impression that it was advocating the constitutional causes, while at the same time fortifying its regulatory grip on the city's nightlife. To prevent this repressive law from being approved by City Council, the NYNA mobilized taxi, hotel as well as restaurant and bar business associations (such as TUFF) to fight to take down the new proposal. The mayor and the DCA continued to contend that the new proposal would not impose requirements as repressive as the NYNA claimed it would. LDNYC leaders also argued that the NYNA was overreacting to the new proposal, and circulating hyperbole about the costs that the new nightlife license would cause (Steinhauer 2004). However, eventually the Mayor's Office lost the public relations battle. The mayor was already in a battle with the city's nightlife industry over his smoking ban, and with the election for his second term near at hand, he did not want to invite another battle. Zoning regulation was the last issue that any politician wanted to face in an election. Hence, the proposal for the new nightlife license was dropped.

As a result of this process, the NYNA realized the power and necessity of making alliances with other sectors to more effectively fight off the City. While the NYNA was originally focusing on recruiting mostly medium to large sized clubs and bars in the city (interview, NYNA representative, January14, 2005), it now started to recruit other sectors of the industry, such as members of TUFF. The industry, marked by fierce competition among individual entrepreneurs, had to transform itself into one in which members collaborated on imminent issues and compromised with each other simply in order to create an effective counter-force to the regulatory regime. The NYNA, while making innuendo that LDNYC had been 'used' by commissioner Dykstra, has also sought to ally itself with LDNYC. In March, 2004, it made two proposals to LDNYC at a meeting (LDNYC listserve, March 3, 2004): first, an immediate action to pressure the municipality (1) to allow 'incidental dancing' in businesses that did not have a cabaret license, and (2) to initiate a sunset provision to completely eliminate the cabaret law's social dancing regulations at some future date, approximately two to four years, in order to protect the investment made by businesses in acquiring current cabaret licenses. The president of the NYNA said that the sunset provision would not need to apply to small spaces, for example, with a capacity of two hundred or less. LDNYC rejected the two proposals from NYNA. LDNYC was against the incidental dancing clause and the sunset provision because the former would still posit dancing as a crime (by not liberating it entirely), and the latter would mean that it would be years before social dancing would be liberated from zoning regulations.

After the new Nightlife License proposal was dropped, LDNYC tried to find a way forward with the object of moving the CPC to work on amending cabaret law zoning regulations, but it was a challenging task. One of the LDNYC leaders said:

> [Members of LDNYC] never got close enough to the DCP [Department of City Planning]. We were told from all angles that zoning is virtually impossible to change—all our advisors told us it'd be a losing battle. (email correspondence, Adam Shore, August 14, 2006)

As was mentioned previously, the CPC and residential communities that were forming strong opinion-setter groups through the mechanism of the CBs, which had become authorized to intervene in land-use related decisions and also propose land-use plans (see Chapter 2), regarded businesses related to social dancing as having much greater land-use impacts than other nightlife businesses, however small they may be. Campaigning to modify zoning regulations in order to deregulate dancing, therefore, would have put LDNYC in an almost endless, ultimately losing battle with the CPC and residential communities. This speaks to how challenging the regulatory system—especially zoning system—has been a daunting project for politically marginalized groups whose claims stand on grounds other than public health, well being, and the quality of life of local actors, the issues that the city has regarded as important to the city's economy and society.

After having failed to find a way to convince the CPC to change the regulations, LDNYC's activism confronted a political cul-de-sac, and faded out by the end of 2004.[16] LDNYC was initially in a favorable position to take advantage of the legitimation crisis of the city's regulatory regime, but it seemed that the regime of interests deeply vested in the city's land-use system was far from vulnerable to the challenge of an organization like LDNYC, especially when it was not accompanied by the support of the NYNA. After the LDNYC dissolved, after little success with its agenda, anti-activists started to consider going to court.

EXPANDING GENTRIFICATION AND NIGHTLIFE REGULATIONS

While the cabaret law reform was thwarted through the process mentioned above, in June 2004, Mayor Bloomberg in collaboration with the Department of Environmental Protection (DEP) put forward a proposal for the first comprehensive overhaul of the city's Noise Code in 30 years. Under the new code, commercial music sources such as bars, restaurants and dance/music clubs would be subject to stricter regulations, as well as new noise sources, such as construction, air conditioning and air circulating devices, Mister Softee trucks (due to the jingle they play) and barking dogs. The new Code would replace the outdated sections in the previous code with ones that would identify new decibel limits for various noise sources, use the latest acoustic technology and facilitate flexible and reasonable enforcement" (Hu 2005) in order to curb noise nuisances that newly emerged for the last three decades due to the increasingly prominent mixed-use character of the city's neighborhoods (Fecht 2004). As one such measure, the DEP suggested replacing the existing code's use of vague phrases, such as "unreasonable to a person of normal sensitivities" with a "plainly audible" standard in certain areas, like nightclubs. The NYNA objected to this provision which would authorize a police officer to write violations for "plainly audible" noises without using a noise meter, as being "too subjective" (Hu 2005). Due to opposition from the NYNA (as well as from Mister Softee executives and construction union officials), the proposal stalled in Council for more than a year.

In December 2005, however, the final version of the new code was approved unanimously by the City Council (for the revised Noise Code, see the NYC government website at http://www.nyc.gov/html/dep/html/noise/index.shtml). As a concession to the NYNA, the DEP removed the proposed provision of inspecting "plainly audible" noise. Instead, in the new Noise Code, officers would be required to carry a noise meter and would issue violations only when it was at or above ten decibels when measuring ambient sound between 7 am and 10 pm, and seven decibels during the rest of the day (a time span including late at night, it should be noted), measured at any point within a receiving property or as measured at a distance of fifteen feet or more from the source on a public right-of-way. In addition, the DEP would be given the authority to waive the $3,200 fine for the first violation if nightlife business owners demonstrated that they would be taking action to reduce the noise (e.g. installing sound-reducing measures). Nightlife businesses would also have to keep their sound level below forty-two decibels, instead of the current forty-five decibels, as measured inside a nearby residence. The commissioner said that the difference between forty-two and forty-five decibels was like "sitting next to someone who goes from silence" to "speaking loudly on a speaker phone" (Hu 2005). The new Noise Code would also restrict bass sounds and their accompanying vibrations and thumps. While the NYNA's representative said that the final version of the code was more or less a mutually beneficial compromise (ibid.), the new code would bring more financial burdens upon under-financed music and dance venues in terms of further sound-proofing. The new Noise Code took effect in July 2007.

In 2006, State Assembly Speaker Sheldon Silver introduced a bill that would eliminate exceptions to the 500-foot rule, but failed to get it passed in the State Legislature. Outer boroughs of NYC opposed the bill as legislators, nightlife industries and real-estate industries there would see an opportunity to attract nightlife to their own boroughs as an urban growth scheme (Ocejo 2009: 195). On the other hand, in the same year, the SLA created a Rapid Enforcement Unit (REU) to streamline the enforcement capacity to effectively regulate "bad" operators of liquor license businesses (ibid.: 204). In 2006, the City Council passed a law that mandated a tightening of existing requirements under state law for the licensing of security guards, including background checks and training for bouncers, following a murder case involving the conviction of a bar bouncer. While at the municipal level new regulatory initiatives continued to emerge, at the neighborhood level the cycle of "gentrification with and against nightlife" has been replicated. Williamsburg in Brooklyn, which was known for the alternative club scene and the presence of artists who had moved out of gentrifying Manhattan in the early 2000s, became gentrified (Zukin 2009), with Williamsburg-based artists and clubs subsequently packing up to move to Dumbo, also in Brooklyn. In 2005, the rezoning of parts of West Chelsea (and the neighboring Ladies Mile and Meatpacking District) was approved, which would open the way to more residential developments in this neighborhood. After the 1990 rezoning of the cabaret law, these neighborhoods were some of the few areas left as-of-right for cabarets. These neighborhoods are currently where the city's destination dance clubs and other nightlife businesses are located, and where, as was mentioned in Chapter 6, the complaints over nightlife businesses have increased among residents (see Figures 7.3 and 7.4). Under the rezoning of West Chelsea, cabarets and other nightlife businesses would still be as-of-right in these neighborhoods, but the move still flew in the face of nightlife businesses in this area, as it was crystal-clear to them and even to the current residents that conflicts over nightlife would mushroom when the area took on a more residential character, and would invite stepped up nightlife regulations (interview, Ivan Schonfeld, a staff planner in the CPC, personal communication, March 30, 2005). Again, nightlife is partly responsible for triggering gentrification in this area, leading a nightlife commentator to call those engaged in nightlife to be "victims of their own success" (Hennessy 2010; also see Zukin 2009: 26). Indeed, it was very soon reported that dance clubs in this neighborhood were remaking themselves into lounge- or restaurant-type businesses, in order to avoid future conflicts with residential neighbors (Ferris 2007).

Rezoning was not the only process through which nightlife has been losing its spatial ground. CBGB, the city's landmark club, was closed for good in the city in 2006, due to the rising rents and disputes over its rent arrears with the landlord, the Bowery Residents Committee (a social service, housing, and drug treatment agency). CBGB sought to receive institutional support from the city (i.e. by being granted a "historic landmark" status), but failed. In 2007, the premises that CBGB had vacated was occupied by a boutique store for world-renowned men's clothing designer, John Varvatos, whose clothing styles often resonate with rock 'n' roll cool (Ocejo 2009: 72). The "symbolic inclusion" of downtown cool (cf. Mele 2000) continued

Figure 7.3 West Chelsea's busy nightscape in August 2011. Photo courtesy of Everita Silina.

Figure 7.4 Gentrification in West Chelsea in August 2011, represented in the InterActiveCorp headquarters building designed by Frank Gehry (on the left), and a condo building located near the newly developed High Line park. Photo courtesy of Everita Silina.

to find its incarnations in gentrifying neighborhoods. CBGB owner, the late Hilly Kristal, said that he would open the new CBGB in Las Vegas where the state seeks to promote nightlife (interview, March 3, 2005). CBGB was not the only entity that sought for cities where nightlife was welcomed. NY-based artists were also leaving for Berlin, Germany—what was now called "the next Brooklyn" (Romano 2004)—where public interests are better distributed to arts and cultures, and there is social consensus on the necessity of protection of small, under-financed art/cultural space, unlike NYC where arts and cultures are encouraged or discouraged mainly according to their market value (interview, A performance club owner in Dumbo, personal communication, December 8, 2009).

AGAINST MARKET AS WELL AS INSTITUTIONAL DISCIPLINE

The NYNA has continued its efforts to contest newly initiated nightlife rules both at the municipal and state level, and to mobilize other nightlife businesses to fight them. It continues to make efforts to establish political representation for the city's nightlife businesses. For example, in 2009, the NYNA also worked as one of the main founding members of the New York Nightlife Preservation Community (NYNPC), with the intention of campaigning for pro-nightlife politicians running in the primaries. The NYNA has accumulated substantial political capital in its relation with politicians and residential communities, but it still remains detached from a range of civil liberty and cultural issues. For example, in 2006, a City Councilmember proposed to require clubs and bars to install surveillance cameras at the entrance and all exits. This proposal was criticized by several gay rights activists and the New York Civil Liberties Union because it would represent an infringement on patrons' privacy (Hu 2006). The NYNA's response to this bill, however, was framed only in terms of the efficiency of using surveillance cameras to secure safety, and not in terms of civil liberties (interview, NYNA representative, December 11, 2009). The NYNA's most important interest is to publicly project its image, as representing "responsible" businesses that respond in a timely manner to the safety concerns of the public. As an association of owners that mostly seeks to procure leverage in favor of their own business interests, the NYNA has an inherent shortcoming as an agent that can promote a more democratic and egalitarian nightlife that would be foundational to fertilizing creative subcultures and subcultural groups.

In comparison, LDNYC advanced a more progressively informed activist agenda, and represented the interests of those (including business owners) who were more interested in the city's subcultural development and expressive interests sheltered within nightlife, such as social dancing. There was no denying the importance of LDNYC's activism, but ultimately LDNYC wound up being a group fighting for a single, circumscribed, issue. The single-issue focus can be seen as inevitable for LDNYC considering the make-up of its membership, and advancing activism with a broader critique of gentrification would have been extremely daunting. As LDNYC leader B reminded us:

> We could not even get this first step [i.e. abolishing social dancing provision of the cabaret law] changed. Just to do what we tried, let alone making substantive changes through the governmental regulatory system, there are just too many players involved—the Department of Consumer Affairs, the Department of Buildings, the City Council, the community boards, the New York Nightlife Association, etc.—and very little cooperation. What we did—major media attention, events and protests, community hearings, etc.—all eventually fell on deaf ears. Those city organizations are the only places that can enact these kinds of changes and they fumbled this over and over . . . (email correspondence, February 21, 2007)

In fighting to claim back a more democratic and egalitarian nightlife, focusing on a single issue would not be sufficient, and may even be counterproductive, as it is easily vulnerable to political maneuvers by governments as we saw from the controversies over the new "nightlife license" proposal.

The political agency that is necessary, therefore, should be the one that fights against a more structural cause, that is, *gentrification* of urban space and the attendant Quality of Life policing in its entirety. Even if LDNYC had succeeded in having the municipality remove social dancing regulations (which it eventually could not), it would have been a limited victory for the city's broader nightlife scene, as the gentrification regime would continue to price out cash-strapped venues, throwing up a repertoire of other regulations to control the "noisy" and "raucous" nightlife. This quandary reminds us of Paul Chevigny's (1991) reflection on the limited repercussions of the court victory (*Chiasson*) of the Musicians' Union in 1988, as was mentioned in Chapter 4. Chevigny observed that even though the legal restriction over the live music venues was abolished after the victory, live music venues were still closing as rents were rising above the level that they could afford. Such history underlines the need of pro-nightlife activism to fight off devastating market forces, such as sharply rising rents, that equally create restrictive subcultural environments for nightlife businesses as well as institutional and legal repressions.

There have been protests by artists over the closure of businesses due to lack of affordability (Carlson 2007). However, these protests have been small-scale and short-lived, with little in terms of more long-term organized mobilization to demand a systematic and sustainable non-market intervention in the market failure over subcultural nightlife venues. Despite the structural difficulties present for small nightlife agents, the NYNA representative opined that a closure of a business because of unaffordability cannot be explained in any other terms than as the incompetence of "individual" entrepreneurs or as shrinking consumer demands in the case of live music businesses (interview, December 11, 2009). This type of reasoning simply shows how far the NYNA stays away from issues relating to struggling businesses and their contingent subcultural communities.

Sometimes, clubs with a long history and popularity become recognized for their cultural worth and are able to generate financial and legal assistance by municipal governments (Moynihan 2009). However, this support is hard to achieve, and cannot be depended on by small, under-funded clubs lacking in history or fame. Therefore, it is important for nightlife actors to demand non-market interventions into the nightlife market, while at the same time fighting off legal and institutional systems that restrict and oppress the creative nightlife environment. The closure of Tonic in 2007, an alternative and experimental music club in Lower East Side, prompted then-Councilmember Alan Gerson to propose a bill for tax benefits for these businesses (Silverman 2007), and also to review the legal validity of legislating subculture/nightlife preservation zones (Office of Councilor Gerson 2007). This proposal was the long-awaited non-market measure proposed to protect subculturally important nightlife venues that have so far been vulnerable to the forces of gentrification and governmental Quality of Life policing. Measures like these would to a significant degree bail out the under-financed, alternative businesses and artists working in these businesses (however, as Gerson failed to be re-elected in 2009, the discussion over this proposal has stalled). In addition, nightlife actors should engage in a more vigorous "war of ideas," campaigning to raise popular awareness of ("noisy" and "raucous") nightlife as an important cultural/social institution. As one of the LDNYC members opined above, it is a hard task to campaign for "noisy" and "raucous" nightlife to be protected from gentrification, as they are stigmatized as anathema to public interests. Nightlife actors should make a claim that nightlife should be noisy and raucous, and that therefore, institutional measures—such as nightlife zones that are not vulnerable to gentrification—are needed to preserve this important cultural verve in the city.

There are other thorny issues that nightlife actors need also to grapple with. The process of writing Gerson's bill (mentioned above) was dominated by live music club actors, with the exclusion of small dance clubs, showing the tensions that have built up between these two contingents.[17] This type of division may be in the way of forming a consistent and united front of pro-nightlife activism. In addition, nightlife, as pioneers and facilitators of gentrification, often directly or indirectly are responsible for social displacement; that is, the displacement of the working-class residences and/or manufacturing land-uses that existed before nightlife moved in together with gentrification (Ocejo 2009). Nightlife often benefits from gentrification, as the latter ushers in potential patrons, and therefore, nightlife tends to concentrate in gentrifying neighborhoods. Subcultural musicians and other creatives—even alternative and experimental ones—often have disassociated themselves from the displaced populations and land-uses, whatever their involvement in this displacement might have been, as Chapter 4 showed based on Mele's (2000) work on the East Village's transformation. This is another issue that nightlife activism should be reflexive about, especially when it seeks to situate itself as part of a broader movement that aims to create democratic and egalitarian urban space.

ACKNOWLEDGMENTS

An earlier version of this chapter was published as Hae, L., 2011, "Gentrification and Politicization of Nightlife in New York City," ACME: An International E-Journal for Critical Geographies, 10 (3) and is available at www.acme-journal.org.

8 The *Festa* Ruling, the Right of Social Dancing and the Right to the City

In 2005, a suit was filed to the New York State Supreme Court by New York City (NYC) dancers and dance organizations, represented by attorneys Paul Chevigny and Norman Siegel. In this suit, *John Festa et al. v. New York City Dept. of Consumer Affairs et al.* (2006; from now on, *Festa*, or *Festa v. NYC*), plaintiffs argued that the cabaret law denied individuals the right to exercise their freedom of expression, as embodied in social dancing. The plaintiffs also contended that the cabaret law was arbitrary, capricious and denied them due process of law. In response, the defendants—New York City Department of Consumer Affairs (DCA), New York City Department of Buildings (DOB), the City Planning Commission of NYC (CPC), and the City of New York—filed a motion to dismiss. Initially, State Supreme Court Justice Michael Stallman allowed the case to continue, but in 2006, decided to dismiss the plaintiffs' challenge, granting the defendants a summary judgment. In dismissing the case, the judge concluded that (1) social dancing was not a mode of expression protected by the First Amendment, and that, (2) given the negative impacts of social dancing businesses on neighboring communities, the cabaret law passed the reasonable test for its legitimacy. The plaintiffs appealed to the State's higher courts, but the latter refused to hear the case. The case was terminated in July 2007.

This chapter engages in an analysis of the reasoning behind the *Festa* ruling. It raises questions about the implications that the court's approach to First Amendment protection might have on the increasingly punitive policing of (spaces for) certain activities and associations enforced in NYC, and by implication, other gentrifying cities in which "Quality of Life" and "Zero Tolerance" are now popular mantras. In NYC "undesirable" populations, subcultures, land-uses and activities have become vulnerable targets of Quality of Life and Zero Tolerance policing, especially under the Giuliani administration. This chapter examines how the judicial sphere is implicated in this punitive policing regime, through two specific discussions.

First, I examine the problems associated with the "categorical approach" recently taken by the courts toward First Amendment protections (Cole 1999), in which courts have distinguished between expressive and non-expressive (or social) conduct and association, so as to provide First Amendment protection to expressive ones, but not to non-expressive ones. I demonstrate through an analysis of *Festa* that the categorical approach represents what Mitchell (2005) has termed "judicial anti-urbanism," in the sense that it thwarts popular challenges to gentrification and the Quality of Life policing that discourages (spaces for) activities that are essential to urban life, if they are not recognized as a constitutionally protected expression. Second, I suggest why we need to take seriously what Mitchell (2005: 580) has called "urban rights" in relation to the broader movement of the "right to the city," which has fought against gentrification and Quality of Life policing. As discussed in Chapter 1, I define "urban rights" as more concrete sets of rights that should be claimed within the broader "right to the city" movements, ones which constitute basic conditions necessary to achieve the "normative ideals" of "the urban" (Lefebvre 1996; Young 1990). These conditions are not easily secured through rights claims within the framework of liberal legalism. I argue that our right to (spaces for) social dancing, if not constitutionally protected, should be valued and protected, as social dancing grants people with opportunities to create and access diverse expressive cultures and to enjoy diverse social interaction and community-building fundamental to ideal urban life. If not a central subject of in-depth discussion in this chapter, I contend that nightlife should also warrant protection on the same ground, as I have argued throughout the book.

I further argue that protecting people's rights to engage in mundane, but fundamental urban activities and rights to access to spaces for these activities from gentrification and groundless governmental punitive policing is a basic step to establish conditions in which we can create more democratic and less commodified geographies of subcultures, creativity, socialization and play. Thereby, the case that this chapter analyzes may primarily focus on social dancing, but the ramifications clearly resonate with numerous urban activisms that have sought to create geographies alternative to those of private property and gentrification (cf. Blomley 2004). In the following, I analyze the *Festa* reasoning in the matters relating to the First Amendment, due process and the "categorical approach." Based on this analysis, I suggest judicial and extra-judicial reforms that the "right to the city" movements may pursue in order to create alternative geographies.

FESTA REASONING

First Amendment and Social Dancing

The *Festa* plaintiffs argued in their legal briefs that there had been few legal cases directly engaged with the issue social dancing in terms of freedom of expression. According to the plaintiffs, *City of Dallas v. Stanglin* (1989; from now on, *Stanglin*), the most important precedent for lawsuits over social dancing, was not really concerned with social dancing *per se* (*Festa* plaintiffs 2005a: 7). The case involved an owner of a dance hall, Charles

M. Stanglin, who challenged the City of Dallas's ordinance that disallowed adolescents from sharing a dance hall with adults. Stanglin argued that the ordinance was an infringement on the adolescent dancers of their freedom of association. In response, the Supreme Court ruled in *Stanglin* that minors dancing together with adults does not "involve the sort of expressive association that the First Amendment has been held to protect" (*Stanglin* 1989: 24). The Court went on to say:

> It is possible to find some kernel of expression in almost every activity a person undertakes—for example, walking down the street or meeting one's friends at a shopping mall—but such a kernel is not sufficient to bring the activity within the protection of the First Amendment. We think the activity of these dance-hall patrons—coming together to engage in recreational dancing—is not protected by the First Amendment. (*Stanglin* 1989: 25)

For the judge presiding over the *Stanglin* case, the primary purpose of social dancing is recreation, and is as devoid of expressive elements in the constitutional sense as "walking down the street or meeting one's friends at a shopping mall." Therefore, the association that is formed for the purpose of social dancing does not warrant the First Amendment's protection.

Festa plaintiffs argued that in *Stanglin*, social dancing was not thoroughly investigated, thus leaving it defined as "recreation," devoid of expressive interests. This was because, they argued, over the course of the *Stanglin* case, the court did not have a chance to listen to the reasoning of the dancers themselves about how they felt about social dancing, and whether they felt it to be a form of expression, as the plaintiff in *Stanglin* was the entrepreneur of the establishment, not its patrons. It was possible that the Court viewed the dispute over social dancing primarily as a conflict over entrepreneurial interest as opposed to as a question of a civil right related to the freedom of expression. Furthermore, as the *Stanglin* case involved minors, the issue of protecting minors may have overshadowed any urgency to examine social dancing as expression (*Festa* plaintiffs 2005a: 8–9). The *Festa* plaintiffs (2005a: 7–8) argued that *Stanglin* became a precedent for subsequent cases that have involved social dancing despite these problems, and that the Court has made little attempt to re-investigate the expressive and communicative value implicated in social dancing. They submitted that *Festa* would be the first case to be put forward to the court which would address the question of whether conflicts involving social dancing ought to be considered as straightforward cases of freedom of expression.[1]

In the complaint, briefs and other affidavits, the *Festa* plaintiffs contended that social dancing deserved constitutional protection as expression. They appealed to freedom of expression under state law instead of federal law, because they believed the scope for protection of expression under the State Constitution of New York was much broader than that afforded by the US Constitution (*Festa* Plaintiffs 2005a: 8). The plaintiffs contended that social dancing deserves protection as an expression guaranteed under the State Constitution for the following reasons: that it is an organic progenitor and an incubator of constitutionally protected expressive arts, such as performance dance and jazz music written for social dancing; that it is an independent art form in itself, "dancing done for aesthetic and communicative pleasure of the dancers, with incidental benefit to those watching" (*Festa v. NYC* 2006: 9); that social dancing "offers a unique mode of expression and communication that is culturally important" (*Festa* plaintiffs 2005a: 11); and that social dancers are engaged in non-verbal expression and communication through movement and who, thus, create a social ritual in which a cultural identity is formed, enjoyed and celebrated (*Festa v. NYC* 2006: 7). US Courts have differentiated between social dancing and performance dancing, recognizing only the latter as a constitutionally protected expression (Hatch 2002); but as Chevigny (2004: 6) has pointed out, these two are both expression, and only "expressively" different.

The *Festa* judge, Justice Stallman, responded to the complaint, first by saying that the free speech provision in New York State's Constitution appeared as part of the United States Constitution, so that "any interpretation of New York's free speech clause should thus begin with a discussion of its federal antecedent" (*Festa v. NYC* 2006: 4). Then, he went on to cite other federal cases before and after *Stanglin*, in which the courts characterized social dancing as not being sufficiently expressive: for example, *Barnes v. Glen Theater* (1991) in which the judge equated social dancing to aerobic exercise. Justice Stallman then examined whether the plaintiffs had successfully demonstrated that social dancing consisted of more than activities and association formed primarily for recreation, contrary to the federal precedents' interpretations. Concurring with *Festa* defendants (2005a: 5–6), the judge said that the plaintiffs' characterization of social dancing and their claim of the social and cultural importance of social dancing did not meet the standards of what the court would understand as "expressive" and "communicative." That is, it was evident to the judge that the "plaintiffs are not claiming that the social dancing that they engage in is *primarily intended* to inspire musicians and choreographers, to assert a cultural identity, or to engage in a ritual with an intended purpose, such as making rain or promoting fertility" (*Festa v. NYC* 2006: 9, emphasis added). The level of the intentionality for communication and expression which would otherwise enable social dancing to qualify as expression was, for the judge, not sufficient.[2]

Justice Stallman then recited the rulings in *Stanglin* and *Barnes* that held that every conduct is in some sense expressive, but that this did not grant every instance of conduct First Amendment protection because "calling all voluntary activity expressive would reduce the concept of expression to the point of the meaningless" (*Barnes* 1991: 581; quoted in *Festa v. NYC* 2006: 10). If this is the case, the judge asked: have the plaintiffs projected any "consistent, practical framework that would classify social dancing as expressive conduct while excluding other physical, athletic, or recreational activities that are arguably similar to social dancing?" (*Festa v. NYC* 2006: 10). The judge continued to question, if, as the plaintiffs proposed, social dancing is expressive given the "aesthetic and communicative pleasure" among its participants, how is social dancing different from, say, "nonprofessional participatory group sports that involve a high degree of skill and social interaction, and give pleasure as much from good form, and from how the game is played, as from victory or the thrill of competition?" (ibid.: 10). Maintaining that the plaintiffs did not answer these questions, the judge concluded that the plaintiffs could not prove that social dancing was a form of expression. The judge briefly added that the degree of expressive interests among different forms of social dancing must vary, but classifying different social dancing according to those degrees would not be what the court should or could do.

Due Process Reasoning

In addition to being a violation of freedom of expression rights, the *Festa* plaintiffs (2005a: 6) contended that cabaret law regulation and enforcement violated due process: that while the NYC government alleged that its regulation of social dancing was directed toward the control of noise and crowding, "the regulation of social dancing by the city's cabaret laws is the regulation of dancing, pure and simple, without some other regulatory interest covering it." They argued that enforcement of the cabaret law was thus arbitrary and capricious, depriving them of the due process required of law. The test for due process, however, differs in cases of violations of First Amendment rights, and cases where these rights are not at stake. For the plaintiffs, social dancing was an expression, so they claimed that the cabaret law should be put under strict scrutiny for due process violation (*Festa* plaintiffs 2005a: 16–18). However, once the judge decided that social dancing did not count as expression, he would only request the relaxed test that was required for the cabaret law's legitimacy in terms of due process.

The plaintiffs raised the possibility that a puritanical prejudice against dancing may be informing the sweeping regulation of social dancing (ibid.: 17–18). The judge unequivocally rejected this accusation, citing that "a law will not be struck down on the basis of an alleged illicit legislative motive" (*Festa v. NYC* 2006: 13), and instead maintained that what should be tested is whether the cabaret law regulations bore a reasonable relation to "the protection of the health and safety of the people of New York City" (which would make the law pass the test of due process). The judge first concluded, invoking the incident of *Happy Land* fire in 1990 that incurred the calamitous casualty in 1990 (see Chapter 6), that the cabaret licensing system, while it imposed additional burdens of compliance, such as safety measures, on businesses allowing social dancing, was not arbitrary and capricious, but a legitimate governmental act to secure public safety. Justice Stallman argued that a business hosting participatory social dancing had an obligation to seriously consider the configuration of internal structures (such as height differential and lighting conditions), as mandated by the cabaret law, in order to protect people sitting at tables (ibid.: 14).[3]

This argument is reasonable, and is not something that the plaintiffs would disagree with. However, the judge's argument leaves unanswered some of the thorny questions that brought the plaintiffs to court in the first place. For example, how can Justice Stallman's point about safety in establishments that have social dancing be addressed to bars where a few patrons drink and/or eat, and occasionally rise up to dance to the music? In addition, should small dance clubs be subject to the same degree of regulatory stringency that applies to mega dance clubs? Was social dancing responsible for the unfortunate disaster at *Happy Land*, or was it that the safety measures that the club was equipped with were inadequate in relation to the size of the club (see Chapter 6)? If the problem of safety arises because of dancers and people who sit at the table sharing the same space in small bars, then, is it not better to create regulatory measures for these types of establishments (again according to the sizes and other features of these venues), instead of imposing sweeping regulations on all sizes of establishments that allow social dancing?

The judge also concurred with the CPC's rationale for the cabaret law's zoning regulations. The regulations mandate that social dancing is off-limits or only permitted when particular requirements are fulfilled in light commercial zones in which residential uses exist closely with light commercial activity, because land-uses for social dancing, like other land uses that belong to Use Group 12 (see Chapter 2), frequently "have a much wider service area, attracting larger numbers of people for varying lengths of time, thereby posing problems of increased pedestrian and vehicular traffic, and resultant congestion as well as increased noise" (*Festa* defendants 2005a: 21). In such zones, only commercial establishments that belong to Use Group 6 are allowed, whose functions are limited to catering to the needs of residential neighborhoods, including small eating or drinking establishments, which can include entertainment but not social dancing, and which can only have a capacity of 200 persons or less. The CPC reiterated its stance in one of the defendants' affidavits (ibid.: 19–20) that if dancing is offered in a restaurant, bar or grill, that establishment is no longer a neighborhood restaurant, and thus would change the character of the land-use in the area from one being primarily residential to being primarily commercial. Concurring with this rationale, Justice Stallman states, "whether dancing, as an activity in itself, causes noise and crowding is not the issue. The issue is rather whether the presence of additional people who wish to dance may cause increased noise and congestion in certain places" (*Festa v. NYC* 2006: 20). For the judge, it was self-evident that bars, clubs and restaurants would draw more people, if they become known to offer an amicable environment for dancing, and in that case, there would be a greater likelihood of increased pedestrian and vehicular traffic, and associated noise.

While the possibility of social dancing may make a bar or restaurant more attractive to patrons, would it create more noise, crowd and traffic if patrons were allowed to boogie to music in eating and drinking establishments with a capacity of *less than* 200 people? The plaintiffs in *Festa* said "no."

> [T]here are indeed crowded and noisy establishments . . . but many of them are crowded and noisy when they have no dancing, *because they are large, because they play loud music, or because the crowd is large.* But there is no nexus between dancing and any of those problems. (*Festa* plaintiffs 2005a: 19, emphasis added)

Along these lines, the plaintiffs argued that the regulation should not be so sweeping as to regulate the entire activity of social dancing, but instead, be differentiated according to the size of businesses, i.e. the number of people that an eating and drinking establishment with social dancing can accommodate.

Plaintiffs invoked *Chiasson* to appeal to the judge that businesses that have live music are regulated according to their sizes, and argued that businesses for social dancing should follow the same logic. The judge responded:

> Because music is recognized as expression that is constitutionally protected under the First Amendment, the *Chiasson* court's scrutiny was exacting . . . In contrast, the social dancing here at issue is not recognized as expressive communication . . . Thus, the City does not here need to make a specific evidentiary showing that the licensing requirement bears a reasonable relationship to the public's health and safety. (*Festa v. NYC* 2006: 15)

For Justice Stallman, given that social dancing was not constitutionally protected and thus only less rigorous standards of due process applied, and given the problems that spaces for social dancing have caused in residential neighborhoods, the rationales for the cabaret law zoning regulations provided by the City to the court were enough evidence to maintain the constitutionality of the cabaret law regulation. He further added that zoning regulations were the results of a carefully considered land use plan produced through a long investigation that "should not be lightly disturbed based on the limited concerns of one group of litigants . . ." (*Festa v. NYC* 2006: 21).

In conclusion, Justice Stallman declared that plaintiffs were not able to rebut the constitutionality of the city's cabaret law, and granted a summary judgment dismissing the plaintiffs' motion. The case was over even before the plaintiffs and defendants could meet in court for more in-depth debates. Unfortunately, the court, which the anti-cabaret law activists turned to as the final resort, did not have, or perhaps more precisely refused to provide, a frame through which the value of social dancing could be articulated, claimed and salvaged from governmental regulations that were based on dubious assertions about (spaces for) any size of social dancing. With this decision, social dancing is still illegal in NYC in unlicensed places, even if the cabaret law is not longer being enforced with the same stringency as it has been at certain points in the past. No Dancing signs are still occasionally found in some nightlife businesses (see Figure 8.1).

Figure 8.1 "No Dancing" sign with graffiti in an unidentified bar in 2005. Photo courtesy of Scott Kidder.

THE CATEGORICAL APPROACH

Although *Festa* may be a matter consisting of the "limited concerns of one group of litigants," the way in which the right to social dancing has been judicially disputed has an important bearing on other legal disputes. It is related to what David Cole (1999) has called a "categorical approach," in which courts recognize more clearly the difference between expressive and non-expressive (or social) associations, and correspondingly between expressive and non-expressive (or social) conduct, particularly since the *Stanglin* ruling (Chevigny 2004: 12). In this approach, association or conduct is constitutionally protected if it is recognized as having an "expressive" interest, whereas ones primarily for a "social" purpose and lacking in "expressive" interests are not. For instance, in *City of Chicago v Morales* (1999, from now on *Morales*), the court rejected a challenge to a Chicago gang loitering ordinance that prohibited associating with and between gang members in public spaces, on the basis that there is no constitutional right of social association.[4]

The line between the expressive and the non-expressive, however, is obviously hard to discern because all conducts and associations are potentially expressive and communicative, as was noted in the *Stanglin* decision, even in the sense of what the court defines as expressive and communicative. Why would it not be sufficiently expressive to walk down the street or to meet friends in the shopping mall? Isn't it that "speech *is* conduct, and actions speak" (Henkin 1968: 79; cited in Cole 1999: 214–15, emphasis in original)? Why would it not be the case that social dancers, or members of a sports team, are intent on expressing and communicating a message to each other, or to observers? If all sorts of arts are recognized as communicative, like "the unquestionably shielded painting of Jackson Pollock, music of Arnold Schonberg, or Jabberwocky verse of Lewis Carroll" (*Festa* plaintiffs 2005b: 13), why can't social dancing be considered communicative? Is the threshold clear? And, as both Chevigny (2004) and Cole (1999) asks, who has the authority to determine whether social dancing is expressive and communicative or not, if the individual who is engaged in the conduct believes it to be so, or if the individual who belongs to an association believes that he/she is engaged in the associational conduct to express something to members of the association or to non-members?

Inevitably, in the context of legal rulings, drawing a line between expression and non-expression has not been without controversy, as such distinctions lacked coherence and as they have been made according to the discretion of individual judges in each case. More problematic has been that "the courts will choose one rubric or the other according as they *want* to support or discourage a social activity" (Chevigny 2004: 14, emphasis added). Puritan prejudice against social dancing (ibid.), or assumptions that connect (spaces of) social dancing to criminality, implicitly prevalent among judges, may explain why courts have denied that social dancing is expressive, as such decisions can exempt municipal governments from being subject to stringent tests for the constitutionality of their regulations, as occurred in the *Festa* ruling. A similar argument has been made in relation to courts' approach to pornography. In her discussion of the municipal campaign to zone out porno shops in NYC, Papayanis (2000: 344) points out that the Court seems to argue that "some forms of protected speech [in this case, pornography] are deemed less worthy of protection than others." Here, too, why pornography is a less protected expression may reflect social taboos about public displays of pornography, and a tendency to associate pornography with some types of criminality.

Chevigny (2004) points out that in previous years, courts took a different tack when considering First Amendment decisions, particularly in the 1960s. Not only was it the case, at that time, that the approach that separates what is expressive from what is non-expressive was discarded in free speech jurisprudence (Cole 1999: 212), the courts also sometimes granted First Amendment protection to activities and associations that they felt needed to be protected—such as labor unions' provision of legal services to members (Chevigny 2004: 28). Often, courts extended constitutional protection to what present courts may define as social associations, "for reasons that were rooted in the fear that the power to interdict 'social' associations was a power to limit association generally" (ibid.: 14), including expressive associations.

This perception must be correct, because what the present court would recognize as expressive associations may not necessarily begin as such *per se*, but instead as an ordinary association with an un(der)defined expressive aim—e.g. meeting friends in a shopping mall—which may later *evolve* into an association with more explicitly pronounced expressive aims through interactions among members (ibid.: 16), such as those that would bring the association within the ambit of constitutional protection. The same can be said in relation to expressive vs. non-expressive activities. In this sense, Justice Stallman's argument that *Festa* plaintiffs did not prove that they participate in social dancing *in order to* inspire musicians or performance dancers, and that this should deny constitutional protection to social dancing, is defective. It is clear that, while not necessarily intending to do so, social dancing often spontaneously inspires and influences other art forms due to its particular aesthetic expression, communicative pleasure and ritualistic festivity. The social dancing practiced in social groups randomly or semi-randomly formed, therefore, deserves constitutional protections. Unless we allow this randomness and spontaneity to unfold, our society cannot support a wide scope of expressive art forms, activities and expressive associations. As Chapter 3 showed, associations among some black gay groups in NYC in the 1960s and the 1970s that developed a community of collective dance were the primordial form of what was later commercially called "disco."

The judge also pointed out that the plaintiffs' contentions were flawed because they were not claiming that the "social dancing that they engage in is primarily intended . . . to assert a cultural identity, or to engage in a ritual with an intended purpose, such as making rain or promoting fertility." However, the judge's interpretation does not consider the ways in which, as *Festa* plaintiffs implied, pleasure in dancing together in an interactive manner among members of a social group evolves into a ritual in which members continuously create and re-create the cultural identity of the group. While it is certainly the case that people sometimes dance together with an intention to assert a cultural identity among members, or in order to be engaged in a ritual, the reverse is also true; the assertion of cultural identity, or ritualization of certain wishes of the group through social dancing, may be the product and effect of the interactive bodily communication between members through social dancing (Buckland 2002; see also Chapter 3).

Therefore, there are enough grounds for courts to discard the categorical approach and extend constitutional protection to so-called "non-expressive" activities and associations. However, presenting the value of these "non-expressive" "social" activities, simply in relation to the First Amendment's mantras of "expression" and "speech," is limiting as I discussed in the Introduction of this book. While non-expressive associations are important as a seed that conceives future expressive associations, they are also not peripheral in accomplishing vibrant and democratic society. This is especially true for urban societies, whose normative ideals include diverse and vibrant social relations, which are incubated and nurtured through various social activities and associations. In the following section, I want to develop this point by discussing social dancing in relation to "urban rights," and the ways in which changed economic and social geographies in cities have affected the (im)possibilities of such urban rights in our society.

TRANSFORMATION OF URBAN SPACE AND URBAN RIGHTS

Judicial Anti-Urbanism

Mitchell (2005) proposes the notion of "urban rights" in his analysis of *Virginia v. Hicks* (2003, from now on *Hicks*). This was a lawsuit involving Kevin Hicks, a young Black male, who was arrested for trespassing on a street in Richmond, Virginia. This street was once a public street, but was at the time *privately* owned by a public entity that presides over public housing projects, the Richmond Redevelopment and Housing Authority (RRHA). The street became the private property of the RRHA in 1997 when the municipality signed over the deed to streets in and surrounding all city public housing projects to the RRHA. The property rights of the RRHA over streets enabled it to legislate and enforce rules that prohibited trespassing onto these streets by outsiders that had no "legitimate business or social purpose" for being in them (Mitchell 2005: 567). A person who trespassed would be presented with a "barment notice," and would be forbidden to enter any properties in the city owned by the RRHA, violation of which would lead to the arrest of the person for trespassing. Hicks, who was not a resident of Whitecomb Court, one of the RRHA public housing projects, but whose mother and girlfriend were living there, was served with a barment notice twice for trespassing on the streets of Whitecomb Court. He was later arrested for entering the street in violation of the barment notice, when he was delivering diapers to his baby, who was in his girlfriend's custody.

In Mitchell's analysis of *Hicks*, two issues can be related to *Festa* that also raise crucial questions about the changing geography of cities and property relations, and the impact of such changes on urban life. First, Hicks's attorneys argued that the street, even if it was privatized, was still a "public forum" because many people were still using the street as if it were still a public street, and that therefore, the barment policies of the RRHA violated First Amendment rights. That is, prohibiting Hicks from being on the arguably "public" street would have an effect of having people whose racial/sexual/class identity is similar to Hicks inhibit themselves from being in that street, and by implication, this meant seizing from these people a place to exercise important rights, such as freedom of speech, on that street. While this argument was accepted in the lower court, the United States Supreme Court eventually did not accept the argument. The Court argued that freedom of speech was never violated by the RRHA barment policy, as Hicks was not exercising this right, and that the state had a compelling interest in regulating non-protected activities such as being on the streets (Mitchell 2005: 581–82).

Mitchell (ibid.: 566) contends that to be (able to be) present in cities' public spaces is an essential right that enables urban social life, that is, what Lefebvre (1996: 2003) called "the urban" (for a more discussion of this, see Chapter 1). This is so because without this right it is impossible to exercise First Amendment rights, as Hicks' attorneys argued. Mitchell (ibid.: 582) further argues that being able to be present on the public street is the crucial precondition for what Lefebvre (1996) called the "right to the city," since it provides the chance "to be visible as a member of the urban public," and therefore works to enhance a democratic urban public sphere. However, the right to be in a public space—as Mitchell (ibid.: 580) has suggested, as an "urban right"—has no place within the Constitution given that no constitutional clause warrants its protection as a right, unless the action is closely allied with other constitutionally protected rights (e.g. delivering a political message in public space). If an activity is not strongly affiliated with any protected right, as in Hicks's case (i.e. delivering diapers to his baby), it is quite difficult to shield this right from regulation by a governing body, if the governing body, by simply going through the relaxed test of the rationality of its regulations (in contrast to the strict scrutiny that must be passed in the case of regulation imposing on First Amendment rights), can "reasonably" prove to the court that the regulation is necessary to secure public safety and well-being.

The second issue that connects Hicks's case to *Festa* is that in the US Supreme Court's reasoning in *Hicks*, there was a lack of engagement with the problems that have ensued from privatization of what once used to be a public street. The lower court pointed out that the privatization incurred "high social costs" by infringing on people's rights to exercise First Amendment rights by diminishing public spaces where they can exercise them (Mitchell 2005: 579). But, the reasoning that the Supreme Court focused on, in contrast, revolved around Hicks's transgression as a violation of the law, and barely paid attention to the urban changes that made it almost impossible for Hicks not to violate the law in the first place—the privatization of what were once public streets, which invalidated the status of the street as a public forum. The absence of this discussion, Mitchell argues, eventually assists in privileging interests that are deeply entrenched in the increasing privatization of urban space, which has been responsible for the decline of "the urban." Mitchell (ibid.) says that the Court's approach as such can be characterized as "judicial anti-urbanism." The *Festa* case also raises the problematics of "judicial anti-urbanism" in a manner similar to Mitchell's analysis of the *Hicks* case.

Social Dancing As an Urban Right

As discussed in Chapter 1, the "right to the city" (Lefebvre 1996) mandates as its key principles the right to appropriate urban life and space, the right to participate in the production of urban life and space, and the right to diversity (also see Duke 2009; Purcell 2003). Drawing on Young's and Lefebvre's visions, we can suggest a set of basic conditions necessary for the realization of the normative ideals of cities—conditions that may be secured by asserting "urban rights," more concrete forms of the "right to the city." Urban rights are comprised of the mandate of opening up democratic access to urban spaces by multiple groups in cities; democratic and egalitarian ownership of means of production and reproduction in society; protecting diverse cultural lives, the spaces that represent them, and the interactions between these diverse groups; supporting creative activities, spaces of use value, and socialization based on pleasure and play, for which profit-making and commodification are not the framing principle. Securing these rights would bring us organic vibrancy unique to "the urban" and, further, lay the foundation for the progressive potential that defies the suffocating forces of commodification.

Mitchell's analysis of *Hicks* suggests that having well functioning public space—that is, public space that grants equal access to every member of urban society, regardless of an individual's economic, racial, sexual, and other social/ascriptive identity—would constitute one of the fundamental conditions for realizing the ideals of "the urban." On the other hand, if granting to every urban inhabitant democratic access to urban spaces of play and of use value can be claimed as another form of "urban right", I argue that rights to (spaces for) social dancing should also be taken seriously within the framework of this "urban right"—as an important condition that enables "the urban." Spaces for social dancing in NYC, such as dance/music clubs, have been contributing to the city's history of diverse subcultures, vibrant social interactions and dynamic identity politics for decades.

As Judith Lynne Hanna (1979) says: "to dance is human;" it is commonly accepted now that dancing is integral to being human. While different from speech, it is still an important way of expressing and communicating one's inner-self with others in a more bodily manner. Therefore, if speech significantly constitutes being human, and is thus an important right, so is dancing. In addition, dance brings a "holistic sensation of total involvement" of mind and body (Hutson 2000), which is another important experience innate to human beings. Social dancing extends such features of individual dance to the collective level. And, just as dance is innate to every human being, social dance is universal to every society, fulfilling countless essential functions at the societal level (Spencer 1985). For example, every society has certain kinds of dance rituals and festivals that celebrate its history, tradition, and cultural identity. Dance is often mobilized for social control and, at other times, articulates with crude commercialism, but this fact only underscores the effective role that social dancing plays in expressing messages and socializing people.

These are the main points of arguments that *Festa* plaintiffs advanced, but I want to argue that social dancing is also an expression and communication with specific temporal (nighttime) and spatial (bars, clubs, etc.) contingencies, that enable a particular method of socialization among citizens unique and central to urban lifestyles and "the urban." Sharing a space for the purpose of music and dance provides citizens a loose, but still exciting, opportunity to associate with strangers, to form a loose group with specific subcultural capital, and to have a collective experience of joy (Chevigny 2004; Hutson 2000). Such socialization focused on social dance often develops into an association very specific to urban society, the function of which has been to create a tribal community among members that reverberated with the vibes of family and church. As was described in Chapter 3, since the late 1960s, such socialization was steeped importantly, if not exclusively, in the cultural politics of youth from racial and sexual minorities in NYC (Fikentscher 2000). Orgiastic celebration and the "carnivalesque" (Bakhtin 1984) that, according to many, have disappeared from civilized western society (Ehrenreich 2007b; Hutson 2000; Reed 1998), were revived in the spaces in which minority communities gathered and celebrated their identities through music and social dancing—spaces like Loft in the 1970s (Chapter 3) and Paradise Garage (Chapter 5) in the 1980s. In a Lefebvrian sense, these spaces symbolized "use value" for this group of people, created through the appropriation of (derelict) parts of the city and through forming their own habitat and *oeuvre*. Such spaces have also become important sites for the evolution of minority politics that play out in conjunction with fun, pleasure, and *la Fête* (cf. Fikentscher 2000; Lawrence 2003). Today, club cultures and social dancing have increasingly become "both a source of extraordinary pleasure and a vital context for the development of personal and social identities" for white youngsters as well (Malbon 1999: 5). Spaces for social dancing also warrant citizens' access to diverse expressive arts because so many (urban) expressive arts—jazz, Latin, rock' n roll, disco, hip-hop, rock, etc.—have been nurtured and developed through spaces for social dancing, as previous chapters evince. The right to social dancing is, therefore, an important urban right that enriches the social and cultural lives of cities.

And, as in the case of some contemporary youth outdoor rave cultures, subcultures premised upon the belief in social dancing and parties as an alternative approach to philosophy, lifestyle, environment and space have also laid the foundation for progressive street politics (Jordan 1998; Klein 2000: 311–24; McKay 1998). In this manner, spaces for social dancing constitute "the politics of difference" (Young 1990)—politics that embody a philosophy different from, and alternative to, existing geographies of social relations in cities. To be sure, social dancing venues are also commercial units, and therefore, labeling them as spaces for alternative politics is in some sense misleading. It is the under-financed clubs, however, that often make important contributions to communities of music, dance and identity politics, rather than affluent ones that focus on profit-making, with music and subcultures as their least concerns, that have been adversely affected by exacting cabaret regulations and expensive real-estate markets. For this reason, the cry out for the right to (spaces for) social dancing (against closure due to market forces as well as excessive state regulation) is important if we are resolved to restore to urban life and spaces organic places that are important in the development of music and dance, and identity-building, and in which market calculus does not predominate.

SALVAGING THE RIGHT TO SPACES FOR SOCIAL DANCING

Like countless mundane activities and associations, social dancing plays important and often radical social, cultural, and political roles in urban society. Therefore, the right to (spaces for) social dancing should be given serious consideration in the face of sweeping governmental regulations that questionably associate *any size of* social dancing with noise, violence and other nuisance. In NYC, the government's primary concern about quality of life has led it to summarily disregard the adverse impact of its sweeping regulations on the city's precious social and cultural fabric. Anti-cabaret law activists turned to court as a last resort to challenge the cabaret law, but the court has taken an approach that resembles "judicial anti-urbanism," by adopting constricted definitions of "expressive" and "communicative," and taking the categorical approach regarding the issue of constitutional protection. The Court's approach allowed the municipality to evade stringent scrutiny of the rationale for its cabaret law regulation and, consequently, the *assumed* criminality of social dancing embedded in the regulatory dictum of the cabaret law was not rigorously tested. Justice Stallman mentioned in his decision that nobody would disagree about "the worth of social dancing" and "the centrality of dance in the development of the arts" (*Festa v. NYC* 2006: 9). Yet, the court did not suggest any ways in which the worth of social dancing, when it is infringed upon by questionable governmental regulations, can be protected. Like the right to be present in public spaces, as in *Hicks*, the right of social dancing, another important right that enables the normative ideal of cities, was not given a chance for serious debate in court, due to its ambiguous standing in terms of constitutional protection. While a range of activities directly related to constitutional freedoms themselves have been susceptible to state suppression, it is more serious in the case of activities that lie at the borderline of constitutional protection, like social dancing, as there are few judicial resources to save them from state suppression, no matter what beneficial roles these activities may play in the cultural and social geography of the city.

It is noteworthy, too, that when unprotected activities, associations or their spaces are regulated, it is quite often particular groups of marginalized people, a particular set of behaviors, and spaces for them that become the main targets and victims of these regulations, even if the regulation may in principle be intended for all citizens. Kevin Hicks's arrest had much to do with the fact that he was a black male, as other "legitimate" looking citizens were freely using the street. Chicago anti-loitering ordinance in *Morales* was also directed to people who hang out with gang members. By not subjecting such regulations to rigorous scrutiny, the court's categorical approach may help to reproduce regulatory practices that are premised on ill-founded assumptions about "undesirable" populations and "disorder." Likewise, not all mundane urban activities may be policed, but activities targeted by puritan moral panics, such as social dancing, more often are, especially activities practiced by gay people or people of color (Buckland 2002).

Missing in the *Festa* decision was also the recognition of the changes in the economic, political, and social geography of NYC—i.e. gentrification—that have rendered even the smallest example of social dancing criminal. While the *Festa* court's reasoning revolved around whether social dancing is an expression worthy of constitutional protection, it does not question, let alone probe into, the gentrification process that has made the city government abuse the provision of the cabaret law in order to keep nightlife businesses in check, and that made alternative, experimental nightlife businesses vulnerable to exorbitant rent markets as well as to Quality of Life policing.[5] It is true that the court cannot do much about increasing gentrification. It is worthwhile, however, to reflect upon the history of judicial matters on similar issues, and note that the court was once ready to extend constitutional protection to social activities and associations that need protection, for good reason. In an era when intensified gentrification and privatization threatens to criminalize (spaces for) various urban activities and associations, the contemporary court seems disposed to abet such trends by adopting the "categorical approach" and by refusing to extend constitutional protection to invaluable urban activities and associations.

The way that social dancing businesses have been contested in NYC for the last three decades, and more specifically, the way in which the reasoning unfolded in the *Festa* decision, may point to directions and strategies for activism both within and outside the judicial sphere to protect important urban activities and spaces through the platform of "urban rights" within broader "right to the city" movements. Most of all, it would involve raising popular awareness of the social, cultural and political centrality of these activities and spaces, the very spaces in which a new urban politicality can be conceived, experimented and lived. Activists should work to raise awareness about how this space has been easily subjected to, and reshaped by, the state's disciplinary tactics—that is, how municipal governments have abused the categorical approach in order to justify Quality of Life policing over (spaces for) important urban activities and associations (the militarization of urban life and space after 9/11 being a good, if deplorable, example of this). This work will be translated into a sense of urgency on the part of the court to extend constitutional protection to these activities and associations and their spaces, regardless of their "expressive" interest. This will provide an opportunity to stringently test the rationality and legitimacy of governmental regulations. Under such strict scrutiny, the cabaret law, for example, would be rigorously tested for questions such as whether the government has a compelling state interest in enforcing the sweeping prohibition of social dancing itself, and whether the law is tailored to further the precisely defined aim of the regulation, through the least restrictive possible means. The effectiveness of cabaret law enforcement in controlling real nightlife problems (noise, heavy traffic, etc.) has been questioned even by some government officials and enforcement agencies. We can assume, therefore, that the case has potential of winning when tried in court.

Furthermore, future lawyers and plaintiffs should push on the limits of the liberal courts further, by bringing broader questions to court and having them debated over in this context. Questions to be articulated should include: given expanding gentrification and privatization, what is the social and cultural cost of Quality of Life policing and the categorical approach taken by the courts? Are citizens left with sufficient spaces for their own creativity, pleasure and joy in the midst of gentrification and privatization of urban space? Within what limited political and judicial environment has anti-cabaret law activism, for example, been working, and what resources can the court provide for such activism?

Fights to reclaim "urban rights" within and through the judicial process, however, constitute a fraction of the broader political movement that urban activists have engaged in. Reform through courts bring about an important, but limited, impact (e.g. no victory in court by anti-cabaret law activism will stop the forces of gentrification that will continue to marginalize under-financed music/dance businesses) and, therefore, the fight to protect "urban rights" should be waged on every level, with judicial reforms as one among many objectives that broader movements seek to achieve. For example, NYC anti-cabaret law activists have continued popular campaigns through, say, street performance protests, to raise popular awareness that social dancing is an inalienable right, and should not be maltreated by authority. Furthermore, movements should fight to pressure the municipal government to provide measures to curb the pernicious repercussions of gentrification and privatization of urban life and space. Citizens can demand a *pro-active* role by governments to secure, against the stifling effects of gentrification, rights to spaces in which citizens can legitimately access diverse urban life, like the proposal made by the former NYC Councilmember, Alan Gerson (see Chapter 7). While I cannot enumerate all of the examples here, political strategies to protect "urban rights" can be variegated, as so many forms of urban activism within the "right to the city" movements have proven to us.

Perhaps more social movements should press to have vibrant and diverse urban life institutionalized as a "right" to which urban inhabitants should be entitled, or even seek to institutionalize the principles of the "right to the city." As a positive right, it would mandate citizens' entitlements to spaces of diverse and vibrant urban life, and would also enable contemporary society to overcome the historical limitations of liberal legalism, by entrenching hitherto unrecognized urban problematics into the legal/judicial sphere. Under such an institutional setting, walking on the street, hanging out with peers in public space or dancing in places that they like would be given institutional protection as basic conditions that enable the "right to the city." The policing of authority over these activities would go through strict scrutiny of legitimacy, and gentrification and privatization of urban space that disable these activities would be rigorously challenged when they interfere with people's rights to spaces of vibrant and diverse urban life.

Admittedly, however, mobilizing people to fight for these rights, to say nothing of institutionalizing these rights, is a highly challenging task, in an era where liberal rights claims—such as property rights—have been normalized in the popular psyche and been empowered to trump other claims. Anti-neoliberal formations have achieved scattered victories across different locales, but have not been able to form a serious political capacity to reverse the neoliberal hegemony, as radical progressive formations have been under neoliberal offensive for almost three decades now. Perhaps that is why we need, all the more, and would benefit from, "the politics of rights" (Scheingold 1974). As I discussed in Chapter 1 of this book, the symbolic and coercive capacities attached to rights may help to engender groundswells of anti-neoliberal (and even anti-capitalist) grassroots mobilizations by shaping a new political consciousness and discontent about the "situation" that we live in. The power of the "right to the city" lies here—the power that encourages people to rise up to "demand the impossible": to demand a radical participatory democracy, economically as well as politically equal and just social relations and spatial system and rich, diverse social/cultural life. This is the power that can enable us to close in on the geographies of *oeuvre* and *la Fête*, not (only) for affluent gentries, but for all ranks of citizens. This is why understanding *Festa* is not simply a matter of concern limited to a specific sector, but to all of us who are keen to create or take back alternative geographies, the geographies of "the urban."

ACKNOWLEDGMENTS

An earlier version of this chapter was published as Hae, L., 2011,"Rights to Spaces for Social Dancing in New York City: A Question of Urban Rights." *Urban Geography* 32 (1): 129–142.

Conclusion

A performance club owner in Dumbo told me that artists are now leaving for Berlin, Germany where public interests are better distributed to arts and cultures (interview, personal communication, December 8, 2009). He voluntarily invoked Florida's (2004) "creative city" thesis to underscore how arts and cultures are important to the competitiveness of cities, and to criticize the municipality's inattention to subculturally important nightlife. On the other hand, Mayor Bloomberg, who is ironically equally inspired by Florida, frequently ensures the public of the city's place as one of the world's foremost cultural and artist centers. In tandem, the City's marketing branch, NYC & Company, boasts of the city's jubilant, fancy nightlife on its website as one of the symbols of the city's attractions. This is the irony that has come with the "nightlife fix" in the city. The governmental as well as academic buzz over creativity and nightlife is actually a recognition that comes only when they are enmeshed with urban economic growth.

Contemporary NYC is characterized by well-known hipster neighborhoods that are filled with bars, restaurants and nightclubs with a distinctively stylish look and feel. But this nightscape now relies less on nightspots that feature DJs, musicians, artists and dancers that are original producers of innovative and experimental cultural/art forms. Instead, it now exists in gentrified forms of nightspots and gentrified zones of cultural commerce, lined up along with upscale residences that primarily cater to professional gentries who can afford the stylish cultural consumption that such nightscapes offer. The city's growth coalition has tried to allure these professional contingents, the so-called "creative classes," whose existence are understood as enhancing the post-industrialization of the city and boosting the city's tax revenues. It is now widely accepted among policy makers that these constituencies are attracted to the cities that can provide cultural amenities and cool urban lifestyles. The transformation of the landscape of contemporary urban societies has been, therefore, very much a phenomenon driven by the culturalization of gentrification, animated by the cultural sensibility and lifestyle aesthetic of a new middle class (Ley 1996), that of the "creative class."

The creative city thesis has gained popularity among various municipalities. Toronto, for instance, where I currently live, is one city which has felt obligated to keep up with the "creative" fad, especially after Richard Florida settled in the city with a position in the Rotman School of Management/Martin Prosperity Institute at the University of Toronto. The former mayor of Toronto, David Miller, who was once affiliated with a social democratic party, declared 2006 to be the Year of Creativity, and promoted various Creative City programs as one of the key planks of his mayoral policy platform. Back in my home country, South Korea, the third largest city, Inchon, has recently trumpeted that the city scored highest in the creative index among the country's cities. Inchon officials have declared that they will further promote the city as the vanguard of creativity in the country. Indeed, cities across the globe are rushing to get on to the "creative" bandwagon, as more cities now feel the imperative to post-industrialize their economies, and to ramp up the competitiveness of their cities within the circuits of the neoliberal global order in which they are now incorporated. Meanwhile, a horde of policy experts have been proselytizing the "creativity fix" (Peck 2007) as the only way to accomplish these imperatives.

What has not been told in all this "creativity" buzz is that the creativity fix has come about with the decline of the very essences that have licensed cities as centers of political, social and cultural creativity. For example, creative artist communities are being forced out of Queen Street West in Toronto due to the recent real-estate boom that has been triggered by the hipster landscape originally created by artist communities there. The condos that are built in this neighborhood will, in all likelihood, be inhabited by the so-called "creative class." In NYC, the subcultural scene—countercultural music, alternative dance communities, young artist communities—has declined with the gentrification that has brought about an infiltration of the creative class, and the post-industrialization that brought in such businesses as media, advertising, design firms, and art businesses. This book has demonstrated this contradiction of the "creativity fix" through the example of nightlife in NYC that has housed creative, alternative subcultural communities. Real-estate values have soared to levels that are unaffordable to most artist communities and alternative cultural nightspots, and institutional supports (including financial subsidies and zoning variances) for these communities and/or establishments for them to stay put have been rare. Equally, zealous protests against nightlife establishments from residential communities—the communities increasingly composed of the creative class—and their incessant calls for a high quality of life has also led to increasing costs in opening and operating nightspots. Rising complaints from residential communities against nightspots have invariably brought about tightened governmental regulations on nightlife venues. In tandem, laws have been expansively legislated and revised—most prominently in the 1990s—as to require nightlife businesses to employ better, and more expensive, schemas of sound-proofing, security, crowd controls, sanitary control of abetting public streets, and in general disciplining the city's nightlife businesses operations such that they are compatible with the imperatives of the quality of life of specific neighborhoods. Community Boards where residential communities wield power have been authorized both by municipal and state governments to shape the size and nature of nightlife in their neighborhoods. In particular, zoning regulations have been so changed as to more easily zone out troublesome nightlife from gentrifying neighborhoods.

The NYC case introduced in this book, thereby, adds to what has been observed among the existing literature on gentrification and displacement of demographics and land-uses that took place in the pre-gentrification period. However, the case examined throughout the book also showed *actually* existing gentrification processes in which constellations of necessary and contingent urban conditions in political, legal, social, cultural and judicial spheres have produced complicated and contradictory patterns of (uneven) development in the fields of nightlife. For example, the case examined in this book showed that the decline of culturally valuable nightspots and the state's suppression of them in the city is ironic, considering the history of gentrification, where these nightspots have played a pioneering role in improving formerly dilapidated neighborhoods. As Chapter 3 showed, discotheques, located in divested neighborhoods of the 1970s and 1980s, helped to breathe life into these neighborhoods, and, were also viable alternatives to landlords of abandoned buildings that were otherwise unleasable for other uses. Dance/music clubs that housed various subcultural communities have also conferred an aesthetic aura and sign value to abandoned neighborhoods, while enhancing the appeal of these neighborhoods for potential yuppie transplants and real-estate capital, resultantly triggering gentrification. This "culturalization of gentrification" was observed in such neighborhoods as SoHo, NoHo, the East Village, Lower East Side, Flatiron District, West Chelsea and Williamsburg, to name just a few examples. But once gentrification kicked in, these neighborhoods saw a groundswell of anti-nightlife rallies among residents, and then a subsequent governmental crackdown on nightlife.

In this process of what I call the "gentrification with and against nightlife," what gradually emerged in the city's nightscape has been the gentrification of nightlife itself, in which upscale/corporate nightspots that are well-financed, profit-oriented and cater to affluent (often young and professional) patrons with expensive bottle service, prevail over under-financed, but culturally important ones. The triple agonies of increasing rental prices, anti-nightlife activism by residential communities, and the municipality's Quality of Life policing have collectively imposed increasingly unbearable costs on these under-financed nightspots, gradually driving these venues out of the city. The gentrification of nightlife has meant that nightlife has experienced "subcultural closure" (Talbot 2006), by which people now have limited access to spaces of diverse subcultures and socialization that have otherwise been enabled by diverse forms of nightlife. While the gentrification of nightlife in NYC has been shaped by various contingent local factors, this complex of processes is one that is recognizable in other locales across the world, bearing resemblance, for instance, with that which has unfolded in the UK. And as UK scholars have argued about the changing nightlife there (Hadfield 2006; Hobbs et al. 2005; Talbot 2006, 2007), the transformation of nightlife in NYC, featuring, too, new patterns of uneven development, has been the product and also facilitator of post-industrialization and gentrification of the city.

This book has also underscored how the municipality has gone to the extent of implementing sweeping regulations that micro-manage routine activities encapsulated within nightlife with serious implications regarding the rights that people are entitled with. With widespread gentrification, privatization of urban space and corporatization of urban entertainments, and the increasing militarization against "undesirable" populations, land-uses and activities (MacLeod 2002), it has been observed that mundane activities—such as walking down the streets, or hanging out with peers in public marketplaces—are unduly regulated in favor of claims to property rights and quality of life. These basic, mundane activities are often not recognized as constitutionally protected rights under liberal legalism (i.e. they are not easily recognized as "speech" or "expression"), which makes it hard for activists to judicially challenge the gentrification and punitive policing that threaten to unduly regulate these activities. City authorities have often been able to go ahead with sweeping regulations over these activities under the rubric of public safety and well-being without fearing that they will be punished for constitutional violations. What these crackdowns on mundane, basic urban activities have often been doing is also to (re-)governmentalize neoliberalized notions of justice, citizenship and the legitimate rights that citizens are entitled to, which prioritize market value and individualism over more egalitarian values.

These mundane activities are foundational in making urban life democratic and open to social and cultural diversity—the normative ideals of urban life (Young 1990)—and this book marks a call that we ought to consider securing these activities through the principle of "urban rights" (cf. Mitchell 2005). "Walking down the street" or "meeting friends in a shopping mall" are everyday activities that people do not think twice about under ordinary circumstances. However, if people are restricted, and even prohibited, from walking down the street because more and more streets are being privatized, or if people who meet friends at shopping malls are subject to numerous codes of behavior enforced by the mall and if there are very limited options to meet friends due to the disappearance of non-commercial public spaces, then claiming the right to walk down the streets, or the right to meet friends without fearing surveillance, become critical issues. Claiming the right to (space for) the routine activities that make up city life has become something that we should struggle for more especially in the context of neoliberalizing and post-industrializing cities where a summary suppression of these activities and spaces has become increasingly routinized.

In this book, I have shown that crackdowns on social dancing—through, among other regulations, the cabaret law that forbids more than three people dancing in unlicensed venues—is not an isolated phenomena, but related to the broader militarization of urban space and life that is characteristic of neoliberal and post-industrial cities. Social dancing is an important "urban right" that enables alternative forms of socialization and identity formation. In the 1960s and 1970s, as Chapter 3 showed, homosexuals and blacks established communities revolving around the themes of dance and music, creating the rituals of particular body politics through which they collectively celebrated their marginality. The role that social dancing plays is not merely a matter confined to history, nor to these particular communities. Spaces for social dancing have been important playgrounds for cultural experimentation, in which various styles, performance and music have been invented and nurtured, and important loci for social interactions for diverse "tribes" among (white) youth (Malbon 1999). Securing the rights to (spaces for) social dancing, therefore, means assuring the existence of basic conditions under which citizens have access to spaces for socializing enabled by bodily communication, and also spaces for producing and consuming diverse subcultures.

The right to (spaces for) social dancing has been arbitrarily dismissed and compromised through the zeal in which the cabaret law has been enforced. I have shown how social dancing has not been recognized as a constitutionally protected form of expression, and how this lack of recognition has authorized the NYC municipality to proceed with a sweeping regulation of spaces for social dancing, with serious implications on the normative ideals of urban life. At the end of the 1980s, after a decade of struggle over discotheques, the municipality amended the cabaret law to drastically reduce the planning zones that would be as-of-right for nightlife businesses that allowed social dance in the venue. More importantly, at that time, the cabaret law was amended in a way that nightlife venues that host *any* size of social dancing would be subject to much more severe zoning regulations than nightspots that offer other kinds of entertainment, especially those entertainments that are constitutionally protected under the First Amendment, such as live music. The provision that imposed much stricter restrictions on any size of social dancing has since then been abused by the Giuliani and Bloomberg administrations, which have found that enforcing the cabaret law as such is an effective way of governing nightlife and of propagating each administration's efforts to raise the quality of life in the city.

Nightlife governance as it was implemented in the city has been a process of struggle all along. In particular, the cabaret law enforcement by these administrations have caused uproars in the city's nightlife sectors and vigorous politicization against the cabaret law enforcement, the context within which the group Legalize Dancing in NYC (LDNYC) has been prominent. Key arguments advanced by LDNYC before it folded, were (1) the assumption that provisions of the cabaret law are premised upon are ungrounded, as social dancing in itself has little to do with nightlife problems, such as increased noise or crowds; and (2) the cabaret law is a grave infringements on citizens' fundamental rights to exercise the sovereignty of their bodies, to enjoy socialization as they wish, and to have access to a variety of cultural spaces in cities. As Chapter 7 showed, LDNYC confronted several challenges in the course of its fight, and these challenges reflected the limited political capital and agency that pro-nightlife actors have been able to exercise within the gentrification regime.

First, the other prominent pro-nightlife lobbying organization, the New York Nightlife Association (NYNA), was not in favor of reforming the social dancing provision of the cabaret law. The NYNA has been described as an organization that represents financially affluent owners of mega dance clubs and upscale lounges and bars that cater to young corporate workers without deep respect for music and dance. Its resistance to reforming the cabaret law has been seen as a move to protect the monopoly interest of these members. The NYNA, on the other hand, sought to fight a series of anti-nightlife legislation, negotiating with the municipality to make pro-nightlife concessions in each piece of legislation. But, as an organization that represents business interests, the NYNA has taken a more regressive gesture when it came to nightlife matters related to civil liberties and subcultural developments, such as the controversies over the cabaret law. The NYNA has often internalized discourses circulating within the pro-gentrification regime about "good, responsible" nightlife neighbors, and even blamed nightlife businesses that did not fit into the image of neighborliness as prescribed by these discourses. This led some nightlife actors to criticize the NYNA as an agent that produces and perpetuates the uneven development within the city's nightlife that has patterned the city's nightlife.

The complicated nature of nightlife enforcement also posed a challenge to LDNYC as a single-issue fighting group. The regulation of social dancing under the cabaret law was merely part of broader crackdowns on nightlife that the pro-gentrification regime has mobilized—that is, crackdowns on nightlife based on rules that regulate noise, security and sanitation, public order and so on. So, when LDNYC sought to liberate social dance from the cabaret law, the anti-nightlife regime sought to make it conditional on accepting more stringent regulation of other nightlife issues, such as noise. Reflecting on this challenge faced by LDNYC, I have made the argument that pro-nightlife activists should develop a more robust and comprehensive challenge to the wide ranging Quality of Life policing and, more fundamentally, to gentrification itself. I have argued that fragmented activisms on single issues have often brought about counter-productive results, and that activisms that only focus on fighting institutional repressions of nightlife and do not advance non-market protection of nightlife can only usher in limited achievements.

The other challenge that LDNYC confronted was the zoning regulations implicated within the cabaret law. It turned out that it was a daunting task for LDNYC to challenge the city's zoning layout, when the latter is closely affiliated with interests of property owners who would be provoked at even modest permissions for land-uses related to social dancing, and when the City Planning Commission (CPC) did not consider that the possibility of liberating social dancing was related to public interests, or at least not sufficiently enough for reform of the cabaret law zoning regulations. Unable to form a united front together with the NYNA in the fight against the city's cabaret law zoning regulation, it was difficult for LDNYC, a group possessed, on its own, with modest political and financial capital, to solve the problem of the cabaret law through political processes. As the history of the city's zoning rule evinces, it is next to impossible for economically marginalized and socially stigmatized groups to move the municipality to change specific provisions of zoning rules.

After having failed in their political efforts to change the cabaret law, anti-cabaret law activists took the cabaret law to the State Supreme Court in 2005, only to hear in the *Festa* ruling that social dancing does not contain the expressive element that would protect it under the First Amendment. This court decision meant that the cultural scene that has constantly produced the creative energy and political edge of NYC would be vulnerable to the economic and political forces that govern them with arbitrary standards, forces that were not subject to rigorous judicial tests of their legitimacy. Like many other liberal judicial approaches, the *Festa* ruling took a very limited and arbitrary definition of "expression" in disqualifying social dancing from constitutional protection, and also refused to question the force of gentrification that has created nightlife problems in the city. While concerned with social dancing specifically, the legal reasoning in *Festa* resonates with the quandaries that have been encountered in the context of other legal and political struggles that have been waged to secure basic "urban rights" in neoliberalizing, post-industrializing and gentrifying cities—such as cross-local movements of rights of democratic access to public space, which has not been captured as a constitutionally protected right in liberal courts.

In this book, I have argued that the "right to the city" can be a radical alternative to the rights frame that undergirds liberal legalism, and showed how this concept can be used as an organizing principle to mobilize people who seek to establish a society of radical democracy in which people's rights to democratically create and appropriate spaces of use value in a more egalitarian way, including spaces for the purposes of play, social dancing and nightlife, are secured. There has been skepticism about framing new movements around the concept of "rights" among scholars who take political economy seriously. I am sympathetic to their critiques, but I have also argued, concurring, for example, with Harvey (2000, 2008), that rights are a pre-condition—a necessary, but not sufficient, condition—for long-term and broader social movements that seek to achieve a democratic and egalitarian urban society. This is because they are instrumental in reshaping popular perceptions, and once institutionalized, they can provide crucial legal legitimacy to activist causes that social movements are advancing.

I have situated the legitimacy of the "right to the city" movements in this context. I have argued that re-imagining spaces for mundane activities, such as social dancing and a range of other nightlife activities within a "right to the city" framework can provide a crucial tool through which to formulate a popular consensus about (1) the importance of (spaces) for these activities, such as social dancing and nightlife, in creating diverse and vibrant urban life, however trivial these activities may seem, (2) democratic participation among diverse urban inhabitants in the production and appropriation of urban space, and more specifically of non-market, use-value oriented spaces of play and nightlife, and (3) the legitimacy to question the legitimacy of the state's sweeping regulations of these important urban activities and spaces. The "right to the city" may also provide a useful frame through which we better understand how the forces of gentrification and privatization, and the regulatory measures that have come with them, have taken us away from the normative ideal of the urban, constituted and enabled by mundane, basic activities and associations, and their spaces, such as spaces for social dancing. Understanding the ways in which the cabaret law was enforced, and the way in which its legitimacy was politically and judicially reasoned, therefore, is not simply a matter of concern limited to a specific sector, as Justice Stallman maintained in the *Festa* ruling. Instead, it can provide an insight to all of us—dancers as well as non-dancers, nightlife revelers or not—who have fought against broader neoliberal offensives against democratic principles and liberal/social rights, have taken the "right to the city" seriously, and are keen to create or take back alternative geographies, the geographies of "the urban."

Appendix 1

Terms of Special Permits for Use Group 6A and 12A before 1990 Rezoning

In C3 Districts, the Board may permit Use Group 6A, for a term not to exceed five years", provided that the following findings are made (NYC Zoning Resolution as of 1981: 350–51):

a) That such #use# will not impair the character or the future use or development of the surrounding #residential# or mixed-use neighborhoods.
b) That such #use# will not cause undue vehicular or pedestrian congestion in local #street#.

The Board may modify the regulations relating to #accessory business signs# in C3 Districts to permit a maximum total #surface area# of 50 square feet of #non-illuminated# or illuminated non-#flashing signs#, provided that any #illuminated sign# shall not be less than 150 feet from the boundary of any Residence District. The Board shall prescribe appropriate controls to minimize adverse effects on the character of the surrounding area, including requirements for shielding of flood lights or adequate screeningIn C2, C3, M1–5A, or M1–5B Districts the Board may permit Use Group 12A, for a term not to exceed five years", provided that the following findings are made (351):

c) That such #use# will not impair the character or the future use or development of the surrounding #residential# or mixed-use neighborhoods.
d) That such #use# will not cause undue vehicular or pedestrian congestion in local #street#.
e) In M1–5A and M1–5B Districts eating and drinking places shall be limited to not more than 5,000 square feet of floor space.
f) In M1–5A and M1–5B districts dancing shall be limited to a clearly defined area of not more than 300 square feet.

The Board may modify the regulations relating to #accessory business signs# in C3 Districts to permit a maximum total #surface area# of 50 square feet of #non-illuminated# or illuminated non-#flashing signs#, provided that any #illuminated sign# shall not be less than 150 feet from the boundary of any Residence District. The Board shall prescribe appropriate controls to minimize adverse effects on the character of the surrounding area, including, but not limited to, location of entrances and operable windows; provision of sound-lock vestibules; specification of acoustical insulation; maximum size of establishment; kinds of amplification of musical instruments or voices; shielding of flood lights; adequate screening; curb cuts, or parking.

Appendix 2
Community Boards in Manhattan

The following shows the neighborhoods that each CB presides over.

CBs	Neighborhoods
CB 1	Tribeca, Seaport/Civic Center, Financial District, Battery Park City
CB 2	Greenwich Village, West Village, NoHo, SoHo, Lower East Side, Chinatown, Little Italy
CB 3	Tompkins Square, East Village, Lower East Side, Chinatown, Two Bridges
CB 4	Clinton, Chelsea
CB 5	Midtown
CB 6	Stuyvesant Town, Tudor City, Turtle Bay, Peter Cooper Village, Murray Hill, Gramercy Park, Kips Bay, Sutton Place
CB 7	Manhattan Valley, Upper West Side, and Lincoln Square
CB 8	Upper East Side, Lenox Hill, Yorkville, and Roosevelt Island
CB 9	Hamilton Heights, Manhattanville, Morningside Heights, and West Harlem
CB 10	Central Harlem
CB 11	East Harlem
CB 12	Inwood and Washington Heights

Appendix 3

The Requirements for Special Permits for Use Group 6C after the 1990 Rezoning (CPC 1989, 40–41)

[T]he Board may permit eating or drinking establishments, with entertainment but not dancing, with a capacity of 200 persons or less for a term not to exceed five years, provided that the following findings are made:

a) That such use will not impair the character or the future use or development of the surrounding residential or mixed-use neighborhood
b) That such use will not cause undue congestion in local streets
c) In M1–5A and M1–5B Districts eating and drinking places shall be limited to not more than 5,000 square feet of floor space
d) That in C1–1, C1–2, C1–3, C1–4, C2–1, C2–2, C2–3, C2–4, C5, M1–5A and M1–5B Districts such use shall take place in a completely enclosed building
e) That the application is made jointly by the owner of the building and the operators of such eating or drinking establishment

The Board may modify the regulations relating to accessory business signs in C3 Districts to permit a maximum total surface area of 50 square feet of non-illuminated or illuminated non-flashing signs, provided that any illuminated sign shall not be less than 150 feet from the boundary of any Residence District. The Board shall prescribe appropriate controls to minimize adverse effects on the character of the surrounding area, including, but not limited to, location of entrances and operable windows; provision of sound-lock vestibules; specification of acoustical insulation; maximum size of establishment; kinds of amplification of musical instruments or voices; shielding of flood lights; adequate screening; curb cuts, or parking.

Appendix 4
Preliminary Proposal for Changing the Cabaret Laws

THE DEPARTMENT OF CONSUMER AFFAIRS, DECEMBER 22, 2003

New York is a vital City composed of ever changing neighborhoods. The diversity and excitement of its lively cultural scene, including nightlife, is an important element of that cultural life. But some poorly managed nightlife establishments cause on-going, quality of life disturbances for their neighbors. In fact, 83% of the City's 97,000 calls this past year to the quality of life hotline were for noise. One week in August 2003 there were 1,972 calls to 311 about noise. The existing laws do not give the City the latitude it needs to effectively regulate problematic places and bring relief to neighborhoods. Working in collaboration with communities, businesses, and other City agencies, DCA proposes to change the law in order to address more effectively some of the quality of life and safety problems caused by poorly managed places without causing an unfair burden on well-run establishments that cause no problems.

Proposal: Eliminate the current regulatory scheme for cabarets and institute a nightlife license instead.

Any public or private establishment located wholly or partially within a residential, C-1, C-2, C-4 or C-6 (1–4) zoning district with a capacity of 75 or more people which is opened after one a.m. (1 a.m.) that wishes to have continuous live or reproduced sound at 90 dBC leq. or higher must obtain a two year nightlife license.

Any public or private establishment located wholly within a commercial (other than C1 or C2), manufacturing or special mixed use zoning district with a capacity of 200 or more people, which is opened after one a.m. (1 a.m.) that wishes to have continuous live or reproduced sound at 90 dBc leq. or higher must obtain a two year nightlife license.

Continuous sound will mean sound lasting not less than one minute.

(Arenas, auditoriums or stadiums that require a special permit from the Department of City Planning and movie theaters are exempt from these regulations.)

The nightlife establishment *will choose* its own maximum dBc leq. level that it wishes to maintain, (as measured 3 feet from any speaker for each separate sound system the establishment operates.

Establishments that meet the criteria less than 3 nights in a calendar year need only obtain a temporary nightlife permit—one per event. The permit shall require proof of all safety permits and certificates, but will not require noise abatement efforts. This permit will cost $120 and such applications must be made 15 days in advance of the event. The applicant will be required to notify the local community board and police precinct of the event by certified mail.

Catering facilities that need a nightlife license will NOT need a catering license, too. But nightlife licensees that hold private functions must follow all the catering rules, as outlined in Section 5–63 of Title 6 of the Rules of New York.

Licensing

DCA will administer the application process, including the solicitation of input from the affected community board, which will have 45 days to report its recommendation regarding the initial application.

DCA will require all initial applicants (and renewals) to submit a current place of assembly permit, certificate of occupancy and, if appropriate, food service establishment permit and liquor license.

The applicant's statements to DCA must be consistent with representations made to other City and State agencies including the NYS Liquor Authority and to the local community board.

The applicant must submit a certified statement from a full member in good standing of the Acoustical Society of America or the Institute of Noise Control Engineers or work for a firm that is a member of the National Council of Acoustical Consultants outlining all necessary measures that the establishment must take to ensure compliance with 241.1(a and b) of the Noise Code at the applicant's declared maximum noise level. A second statement, prepared by the acoustical consultant, must certify that all recommended measures were taken.

The application will include a certified diagram showing the placement of all speakers including those that might be brought in by freelance DJs.

DCA may conduct qualifying and/or compliance inspections to ensure that the nightlife establishment has taken and maintained all the measures recommended in the report prepared by the sound engineer to mitigate the impact of the sound.

Every establishment that has a capacity of more than 499 must have one state-certified security guard for every 50 allowable occupants (plus supervisor) and deploy sufficient guards to maintain order outside the establishment when needed.

Such guards will be required to produce identification and their state license on demand. In addition, if enforcement personnel request a roster of the guards on duty it must be made available immediately. Furthermore, guards must cooperate fully with law enforcement personnel.

Every nightlife establishment must post in a public and visible place a notice that says loud sounds may be harmful to one's hearing, with wording and size determined by the Department of Consumer Affairs.

DCA will review any purported transfers to make sure that they are arm's length transactions from anyone whom DCA previously proved unfit to hold a DCA nightlife license.

Unless approved by the Commissioner, DCA will not issue temporary nightlife licenses.

The license fee will not exceed the cost of administering the license, as determined by Office of Management and Budget.

Enforcement

Every nightlife establishment will provide to DCA, the local precinct and community board the name and number, including the cell phone number, of a responsible staff member of the nightlife establishment, who will be available during hours of operation.

DCA inspectors may inspect licensed nightlife establishments to ensure compliance with all license laws, including the self-certified intended maximum noise level, as stated in the initial application.

DCA will be authorized to write license law violations on 1) exceeding the stated maximum decibel level 2) DEP's Noise Code (241.1 sections a/b) as it relates to the nightlife establishment. These violations of the Noise Code written by DCA will be returnable to the Environmental Control Board, but available for use by DCA. in determining fitness.

(Any and all establishments, whether licensed or not as a nightlife establishment, must obey all relevant aspects of DEP's Noise Code.)

Appropriately trained DCA inspectors will use computerized sound meters that are recognized by the industry as reliable and accurate (Type

1 integrating sound level analyzers in accordance with ANSI S1.4). The meters will be calibrated regularly to maintain accuracy.

Upon demand by a DCA inspector, a licensee must agree not to lower volume levels during an inspection. Failure to comply shall create a rebuttable presumption that the noise level exceeded the self-certified maximum and DCA can write a license law violation.

DCA inspectors will be authorized to order occupants to vacate the premises immediately if exit doors or fire doors are found blocked or locked and write license law violations accordingly.

Nightlife establishments will be required to make a good faith effort, using security guards and sidewalk barriers, as needed, to ensure that the crowds entering or leaving do not cause disturbances. DCA inspectors can write violations on failure to use best efforts.

Nightlife establishments must sweep the sidewalk and 18 inches into the street in front of the establishment within 30 minutes after closing time or by 6:00 a.m. whichever comes first. Failure to do so will be a license law violation. Other sanitation violations will NOT be considered a DCA license law violation.

If odors or stains warrant it, the establishment must wash the sidewalk in front of the establishment within 30 minutes after closing or by 6:00 a.m., whichever comes first. Failure to do so will be a license law violation.

Adjudication

The Commissioner or her designee will determine fines for license law violations that will range up to $250 for the first violation, $250–$500 for the second violation, and $500 to $1000 for the third violation.

If an establishment violates DCA license laws (including exceeding the self-certified maximum sound level) three times within two years DCA shall have the power to padlock the establishment for between one and ten days. DCA will NOT have the power to padlock any establishment for violations written by other agencies.

DCA may revoke a nightlife license only if the location is indicted for two of the following on different days within two years. These will include incidents of: Homicide, Assault, Rape or attempted rape, Possession of weapons, Unlicensed sale of liquor, Sale of liquor to minors, Overcapacity, Disabled sprinkler systems, exit signs or emergency lighting, Blocked or locked exits, Two DCA padlocks

DCA will fine any nightlife establishment that operates without a nightlife license $200 per day from the writing of the violation to the hearing and/or resolution of that violation.

In addition to these fines, if DCA finds that a nightlife establishment has operated without a license after having been adjudicated for unlicensed activity, DCA will have the right to immediately padlock that establishment until it pays all fines and either becomes licensed or ceases all unlicensed activity.

Notes

NOTES TO THE INTRODUCTION

1. I define social dancing as collective dancing between more than three people to pre-recorded, DJ-generated or live music.

NOTES TO CHAPTER 1

1. Not everyone was a winner in this arrangement, as the benefits gained by the working classes were not equally distributed among different racial/ethnic and gender groups.
2. The "mode of regulation" vindicates that the state is actually deeply involved in the market, and de-bunks the belief, especially among neoliberals, that market equilibrium alone can move a capitalist system along.
3. David Harvey (1982) shows that when returns to investment in the primary circuit of capital (manufacturing and other productive sectors) fall below those in the secondary circuit (generally, real-estate and land development), capital switches from the former to the latter.
4. Wilson and Kelling (1982: 31) conducted an experimentation in which it is shown that if (the signs of) disorder in a community, like broken windows, are left untended, they invite further disorder. Such disorder signals to outsiders that "no one cares, so breaking more windows costs nothing." Based on this observation, Wilson and Kelling argued that "undesirable," "obstreperous," "unruly" and "disorderly" people, while they committed mostly minor offenses, also invited more serious crime to the neighborhood. Even merely disorderly behavior, thus, required rigorous policing.
5. BIDs (Business Improvement Districts, or Business Improvement Areas as it is called in Canada) are "a geographically continuous areas that vote to assess property owners a special fee in order to provide additional services" (Kohn 2004: 82). In return for paying these fees, owners are granted autonomy in matters of spending (e.g. spending in policing and cleaning of public space located within the BIDs).
6. This move dovetailed well with offers by pubcos (stock exchange listed pub chain companies) to invest in disused buildings in central cities to recycle them into nightlife establishments (Roberts 2006).
7. While scholars who have studied nightlife have commented on differences in nightlife governance between European, Britain and North American cities (for example, see Ocejo, 2009: 17), comparative research on nightlife governance in different localities still awaits more systematic scholarly attention (for an exception, see Hadfield 2009).

8. The importance of rights (and other liberal institutions in general) in socialist struggles has been long debated within the Marxist tradition. The dominant interpretation among Marxists of Karl Marx's approach to the subject, in writings like *The Jewish Question* and *Critique of the Gotha Programme*, is that Marx was critical of rights. This interpretation has not gone unchallenged among other Marxists.
9. This is why Boyd (2009: 585) argues against the ahistorical theorizing of rights often found among some schools of Marxists. They often erroneously apply Marx's critique of rights, which was moored in the realities of the 1800s, to the analysis of contemporary rights. Marx's critique was concerned with the "rights of man"—first-generation civil/political rights—but not with economic/social rights and the third-generation solidarity rights that did not come to life until the mid-1990s through struggles by socialist and diverse civil rights movements. It is politically dangerous to assume that different rights are "internally homogenous," and that rights are to be "accepted or dismissed as a whole" (ibid.). It ignores the historical evolution in the forms and contents of rights, and also leads to miscalculated, foregone conclusions about the potency of rights.
10. The production of "space" in the direction of *oeuvre* and use value was as important for Lefebvre as other political strategies advanced by socialist movements, as reflected in the following quote: "A revolution that does not produce a new space has not realized its full potential: indeed it has failed in that it has not changed life itself" (Lefebvre 1991: 54). This also reflects Lefebvre's affiliation with the Situationist International, for whose members de-marketizing and democratizing situations—including the spaces—of living and working was key to their campaigns.

NOTES TO CHAPTER 2

1. In zoning terminology, "C" represents commercial zones and "M" represents manufacturing zones. Though not listed here, "R" represents residential zones. The higher the number following the letter usually signifies that the zone is closer to its core designated characteristic; that is, C5 has more of a commercial character than C1, which, in turn, has more residential characteristics than C5.
2. C1 and C2 districts are designed to serve local needs, C3 is for waterfront activity, C4 is for shopping centers outside the central business district, C5 and C6 districts are for the central business districts which embrace office, retail, and commercial functions, and lastly, C7 and C8 are for large commercial amusement parks and heavy repair services (DCP 1990: 75).
3. M1–5A and M1–5B refer to loft zoning areas, where industrial buildings were converted to residential use. Examples include SoHo and NoHo districts in Manhattan (DCP 1990: 105).
4. The idea of Community Boards (initially called Community Planning Boards) was initiated by Jane Jacobs in the early 1960s. Jane Jacobs was an urban activist, who fought the paradigm of top-down urban planning processes often manifested in Robert Moses' renewal projects, which demolished the very fabric that was key to urban vibrancy and promoted suburbanization. CBs were designed to promote neighborhood self-governance and decentralized urban planning processes. ULURP was initiated in the mid 1970s to enhance community participation in the city's land-use planning (Forman 2000).
5. The members of each CB are appointed by the Borough President from an applicant pool, half of them from nominees made by the district's City Council member(s) (Forman 2000).

6. This provision was inserted as a clause of the Administrative Code and made into law in 1990 (see Chapter 6).
7. The Cabaret Digital Video Surveillance Cameras Certification was added to § 20–360.2 of the New York City Administrative Code in 2007 (see Chapter 7).

NOTES TO CHAPTER 3

1. In the New York region, small businesses lost 310,000 jobs between 1972 and 1975, two-thirds of this due to a decrease in manufacturing. Tax delinquencies on loft buildings doubled in the years 1973/74 to 1974/75, from $3.5 million to $7.9 million (Zukin 1989: 29–31).
2. The municipality's legalization of artist residence in these neighborhoods eventually helped to price out remaining manufacturing tenants, thus stepping forward to the long planned de-industrialization of these neighborhoods (Zukin 1989).
3. However, the composition of the Loft's patrons was quite mixed in terms of race, ethnicity, gender, sexuality and even class.
4. A *New York Times* article (Thomas Jr. 1974) reported the conflict between Mancuso and his neighbors in a piece written with seemingly high respect for Mancuso. In a warm tone of writing, the write-up listed the hospitality that Mancuso had been offering to his neighbors. In the report, Mancuso was described as sprucing up the façade of the entrance of the building, sound-insulating his apartment to prevent his neighbors from being disturbed by his record player, changing the entrance keys regularly at his own expense and issuing new keys to the other tenants for their security, and inviting his neighbors to his party without charging fees.
5. By this time, the practice of LSD-induced rituals seemed to have disappeared from the Loft due to the governmental ban on the substance (Lawrence 2003).

NOTES FOR CHAPTER 4

1. For example, during the Koch administration, AT &T received $42 million in property tax abatements in exchange for a promise to keep jobs in the city when it threatened to move 1,000 employees from its headquarters in the city to New Jersey (Sleeper 1987: 437).
2. The area where the bar was located was most likely a C6–4 zone surrounded by residential zones in the Murray Hill neighborhood. That C6–4 zone was shown to be zoned down to C1–9 in the 1983 zoning map, indicating a reinforcement of this area's original residential character, a reinforcement related to expanded gentrification.
3. I could not find documents that detail the successful or unsuccessful legislation of this rezoning proposal. However, it seems that this rezoning effort did not come to fruition at that time.
4. The law came to be called the "Blue Angel Law."
5. This bill, it seems, was not passed. I was not able to find a document explaining why this was so.
6. Cabarets could not be located in residential areas. However, public officials tended to refer to light commercial zones as "residential areas," as light commercial zones started to take on a more residential character with gentrification.
7. The inspection remained quite haphazard, however (Chevigny 1991: 118), such that some clubs were not even aware of the tightened regulation until

1986. It may have been the case that the inspections and occasional raids were directed more toward areas and venues that were the main objects of residents' complaints. This will become clearer from the mid-1980s onwards.

8. Chevigny, who was at that time a professor at New York University's law school and also staff attorney at the New York Civil Liberties Union, started to work with Local 802 on the problems of the cabaret law.
9. The neighborhood that is currently called East Village was called the Lower East Side until the later 1980s. Here I refer to this neighborhood as it is currently known—the East Village.
10. In Figure 2.1, this area is mostly zoned as a residential district with light commercial (C1 and C2) overlays along Avenues A, B, C, and 1st, 2nd, and 3rd Avenues. C1 districts were off-limits to cabarets, but in the C2 districts they were allowed to operate legally through special permits. It is also quite possible that during the 1970s, when this area was abandoned, cabarets may have been able to operate in residential and C1 zones of this neighborhood, on account of the relaxed cabaret law enforcement at that time. On the other hand, in the East Village and the Lower East Side, non-conforming uses, such as bars (but not cabarets), were allowed to locate in residential zones of R5, R6 and R7, if the storefront had non-conforming uses at any point in the building's history (Ocejo 2009: 146). Again, the authority made the distinction between businesses without and with "cabaret" entertainment, and applied unduly stricter zoning regulations over the latter, regardless of the sizes of these businesses.
11. One of the ten owners of Peppermint Lounge was charged with racketeering in the Manhattan Federal District Court by the then United States Attorney Rudy Giuliani, who would later become mayor of the city. With the emergence of crack cocaine during the 1980s, it was more frequently observed that some mega dance clubs became sites of drug trafficking and gun violence. The federal and local authorities started to step up the investigation of these mega clubs, and the closure of the Peppermint Lounge was one of the consequences of this move.
12. If the place was not zoned as-of-right or as requiring special permits for cabarets, it was almost impossible to get a zoning variance to convert the current building code of the place to Use Group 12. If a building existed in an abandoned and decaying condition during the mid- to late-1970s, the practice of using it as an unauthorized cabaret was tolerated by the authorities; in the 1980s, when the area was being gentrified, and inspection by authorities was tightening, such practices became increasingly less frequent.
13. If there had been performance only at the club, without liquor being served, the club would have not been subjected to cabaret inspection.
14. This kind of argument may have fed into the city's contention that more musicians would cause bigger crowds. Chevigny argues that while this may be the case, exponential crowding in the neighborhood would not be generated, as long as small sized venues were allowed to locate in residential areas. An engineer also testified in an affidavit that small neighborhood bars in light commercial zones (C1 or C2) could not book big name performers anyway, so the traffic impact of these small bars, even though they had live music, would not be serious (Chevigny 1991: 125).
15. Chevigny's book, however, does not comment on why he did not consider including dancers (instead of dance club owners) as plaintiffs.
16. One realtor testified that they sold a 2,000 square foot duplex loft in the neighborhood with a 1,000 square foot private roof for $900,000, whereas two years prior to this, the same loft was sold for $675,000 (Hawkins 1988). The area was growing rapidly.

17. In 1983, it was reported that a 1,450 square feet loft was priced at $196,000, and a 2,240 square feet one was sold at $341,000 (Landis 1983). In 1989, prices ranged from $151,000 for a studio to $720,000 for a penthouse unit with a roof deck (Noble 1989).

NOTES TO CHAPTER 5

1. The boundary between underground and overground has always been hard to discern in NYC clubland. DJs and artists associated with underground music do not necessarily refuse to enter into the mainstream market, and underground DJs often deejay in tacky, commercial clubs considered uptown (Lawrence 2003). Bearing in mind the blurriness of the boundary between under- and over-ground in mind, we can still define what ought to be categorized as underground clubs. Underground clubs are usually those where clubbers prioritize artistic elements of music and dancing over any other aspect; that feature DJs who demonstrate a tendency to play relatively little-known underground music or fresh, remixed music for their audiences; where creative music, dance and performance are experimented and played with right in the club; that has no dress code; and where drinks and admission costs are relatively cheap. In NYC, the underground dance clubs have also been those that cater to socially and/or economically marginalized sexual and racial groups, whose identity politics have been closely related to the subculture of music and dance (for further description of underground club cultures, see Fikentscher 2000).
2. Such downtown clubs in the 1980s included, in the East Village, Save the Robots, World, Area, Limbo Lounge, Danceteria; and in the Flatiron District, Sound Factory.
3. Garage refers to the music developed at the Paradise Garage by DJ Larry Levan; house music refers to the music developed in the club Warehouse, in Chicago, by a DJ Frankie Knuckles. Larry and Frankie were close friends, and they learned DJ techniques at Loft and Gallery (see Chapter 3).
4. To be sure, this was not a new move. Social dancing had been regulated by itself under the cabaret law before this instance of rezoning (see Chapter 2), and never according to the size of venues or social dancers. What was remarkable about this rezoning was that when the municipality was going to rewrite the cabaret law so that live music would be regulated by the size of the venues, it did not consider applying the same revision to social dancing.
5. Under this proposal, Use Group 12 would have to close by 12:30 a.m. on weekdays and 2:00 a.m. on weekend nights.
6. This represented a slight twist of the administrative language from "incidental" (in the former cabaret regulation, see Chapter 2) to "accessory" music (in the currently proposed regulation).
7. In the initial provisionary proposal in 1988, it was the 150-person capacity that divided Use Group 6 and Use Group 12. In the newer proposal in 1989, this number was changed to 175, apparently due to Local 802's protest that a 150-person capacity was too small.
8. The businesses licensed by the SLA were allowed to stay open until 4:30 a.m. by state law, while the city tried to shut down Limelight at 4:00.
9. McHugh (1989) reported that the CPC's decision to drop the hours restriction was also due in part to the successful mobilization of club owners in voicing their opposition to the first proposal.
10. For the requirements for special permits for Use Group 6C, see Appendix 3.

NOTES FOR CHAPTER 6

1. For this reason, a vibrant bar scene is often taken as a sign that the neighborhood is going through gentrification (Ocejo 2009: 104).
2. In 1989, five people were killed, another eight were injured in stabbings or shootings, and dozens of others—including a group of farmers setting up a stand across the street—were assaulted in or near the club. In early January 1990, three people, including two by-standers, were shot in front of the club.
3. This bill was also called the Padavan Law, after the state senator who initiated it.
4. Other exceptions were also given to "grandfathered" establishments, that is, establishments that were licensed before November 1, 1993, the date that the 500-foot rule took effect, and the establishment that were renewing their licenses.
5. In addition, officials in the SLA are not elected, but appointed by the governor, so they are less concerned with appeasing constituencies (Ocejo 2009: 192).
6. As a matter of fact, George Kelling, one of the originators of the "Broken Windows" thesis, came to NYC to educate high-ranking NYPD personnel (Gotham Gazette 2005).
7. The Police budget rose from $1.7 billion in 1993, to $3.1 billion in 2000 (Mitchell and Beckett 2008: 89).
8. The number of misdemeanor arrests increased from approximately 129,403 in 1993 to 224,668 in 2000, for an 80 percent increase in total. The number of drug arrests also increased, from 66,744 in 1993 to 140,122 in 2000. Marijuana arrests experienced a stunning sixfold rise, from under 10,000 in 1993 to over 60,000 in 2000 (Mitchell and Beckett 2008: 89).
9. It has been argued that the drop in the crime rate, which was also experienced by other major US cities at that time, was related to a wider economic cycle and the decline of crack cocaine-related violence.
10. The share of the city budget allocated to the Department of Social Services declined from 23.8 percent in 1993 to 13.4 percent in 2001, a drop of 43.8 percent (Mitchell and Beckett 2008: 86).
11. For a similar observation, see Ocejo (2009: 295).
12. Buckland (2002: 130), however, argues that among the clubs harassed or closed down, queer spaces, especially queers of color, and women, were overrepresented.

NOTES TO CHAPTER 7

1. It seems that at some point between 2000 and 2002, the Multi-Agency Nightclub Task Force was changed to the Multi Agency Response to Community Hotspots, which was popularly called MARCH. The difference between these two is not clear, but they apparently fulfilled almost all the same functions and they were both supervised by the Deputy Mayor's Office. It is possible that the Bloomberg administration, once in office in 2002, may have simply given the task force a slight name change.
2. The policy platform of Michael Bloomberg, who was a billionaire businessman before running for the mayor, was based on economic pragmatism and neoliberalism. For an analysis of his neoliberal policies, see Brash (2011).
3. For a legal scholar's concurrence with this argument, see Hatch (2002).
4. Councilmember Alan Gerson also went public to argue that the cabaret law is unfair and unreasonable. He said in an interview, "If an establishment is appropriately soundproofed, government should not care if people inside dance. Otherwise, it is an impingement on personal expression" (Burger 2002).

5. All the listserve citations are anonymous.
6. However, the internal discrepancies continued to exist even after this, as some members continued to argue that LDNYC should accept Gerson's proposal (e.g. Anderson 2002).
7. LDNYC argued that nightlife businesses should be regulated according to their size, and not whether they have dancing or not. Therefore, LDNYC was not in agreement with the incidental dancing exception, as it left the core of the dancing provisions of the cabaret law untouched. In addition, as was mentioned in Chapter 5, the incidental dancing can be interpreted by the enforcement agencies in a quite arbitrary manner—e.g. businesses may or may not be fined depending upon, say, whether it is six or seven people dancing. The boundary between incidental and non-incidental, serious dance can be subjective and controversial.
8. "Bottle service" refers to the practice of clubs making it mandatory for patrons to order at least one bottle of whiskey or wine at every table at an exorbitant price. The recent popularity of this practice among clubs in the city reflects that the demographics of club patrons have been changing. In an interview with the *New York Post* (Niemietz 2006), Webster Hall curator said, "[bottle service] caters to Wall Street people who just don't care about spending $5,000 to $6,000 in a night." Bottle service has also arguably been useful to screen out unwanted patrons of the sort who usually are unable to afford the service.
9. The B&T or the "Bridge and Tunnel" crowd has been a popular term in Manhattan, used by local residents and elite downtown nightlife actors to disparagingly refer to suburban residents who come to NYC for its nightlife on weekends using either a bridge or a tunnel.
10. On the other hand, from communications with LDNYC leaders, it seems that LDNYC thought the reform of the cabaret law by the DCA would concurrently involve the reform of the cabaret law zoning system.
11. The Paid Detail Unit (PDU) is a system through which off-duty, uniformed police officers can be hired to provide security services.
12. The NYNA has been citing evidence from New Orleans and Miami Beach that have seen great noise reduction through the use of PDUs by nightlife businesses (also see Berkley and Thayer 2000).
13. For the complete proposal, see Appendix 4.
14. While the zoning resolution of the cabaret law is legislated by the City Planning Commission, which is part of the Department of City Planning (DCP), the officials of the Department of Buildings inspect the violation of the zoning law.
15. Commentators inferred that the DCA may have not been well versed in the city's byzantine zoning code tethered to the cabaret law, and that was why it claimed that with the new nightlife license law, that dancing would be liberated (Brown 2004).
16. Since then, a group called "Metropolis in Motion" was formed and has been hosting performance protests on the city's streets.
17. This tension comes from the perception on the part of music clubs and musicians that dance clubs by far do better in the market than music clubs, and that therefore dance clubs do not need institutional bail-out (interview, New York-based DJ, personal communication, Dec 22, 2009).

NOTES TO CHAPTER 8

1. Lawyers representing *Festa* plaintiffs also sought to recruit club owners, but the latter refused to be part of the lawsuit because they were fearful of possible retaliation from the city government (email correspondence, Paul Chevigny, June 27, 2005).

2. *Festa* defendants (2005b: 6) also argued that, even acknowledging the fact that recreational dancers had the intention of conveying an idea, it was still not the case that "recreational dancing is a form of expression likely to be understood by the viewer as an attempt to convey a particular point of view."
3. While the judge reviewed the plaintiffs' challenge to the cabaret licensing requirements, and responded to it, he also mentioned that *Festa* plaintiffs lacked standing in putting forth this challenge, as it should be business owners that should bring individual cases to court to challenge licensing requirements. As was mentioned, business owners did not participate in any capacity (except Baktun's owner) in *Festa* for fear of the retaliation from the authority.
4. The Court eventually struck down the Chicago anti-gang ordinance, however, reasoning that the ordinance was vague, and could be in danger of being applied to non-gang members.
5. Such recognition was not only missing from the judge's decision, but also in the *Festa* plaintiffs' arguments to the court. The plaintiffs' reasoning was restricted to accusing the government of exercising puritanical prejudice against dance, which they saw as being centrally responsible for the sweeping regulation of spaces for social dancing. See Chapter 7 for my discussion of the disassociation of anti-cabaret law activists from the problematic of gentrification.

References

Aletti, Vince. 1975. "SoHo vs. Disco." *Village Voice*, June 16.

Anderson, Lincoln. 2002. "Let the People Dance! Gerson to Lead on Cabaret Law Reform." *The Villagers*, November 7.

Atkinson, Rowland. 2003. "Domestication By Cappuccino or a Revenge On Urban Space? Control and Empowerment in the Management Of Public Spaces." *Urban Studies* 40 (9):1829–43.

Bagli, Charles. 1991. "'De-gentrification' Can Hit When Boom Goes Bust." *New York Observer*, August 5.

———. 2005. "Largesse." In *America's Mayor: The Hidden History of Rudy Giuliani's New York*, edited by R. Polner, 49–62. Brooklyn, NY: Soft Skull Press.

Bakhtin, Mikhail. M. 1984. *Rabelais and His World*. Translated by H. Iswolsky. 1st Midland book ed. Bloomington, IN: Indiana University Press.

Barbanel, Josh. 1990. "Bronx's Social Club's Sublease: How a Firetrap Skirted the Line." *New York Times*, March 28.

Bartholomew, Amy. 1990. "Should a Marxist Believe in Marx on Rights?" *Socialist Register* 26: 244–64.

Bastone, William. 1997. "Quality of (Night) Life Issues." *Village Voice*, February 18.

Belina, Bernd. 2004. "From Disciplining to Dislocation: Area Bans in Recent Urban Policing in Germay." *European Urban and Regional Studies* 14 (4): 321–36.

Bell, David, and Jon Binnie. 2005. "What's Eating Manchester? Gastro Culture and Urban Regeneration." *Architectural Design* 75 (3): 78–85.

Berkley, Blair J., and John R. Thayer. 2000. "Policing Entertainment Districts." *Policing: An International Journal of Police Strategies & Management* 23 (4): 466–91.

Berman, Marshall. 1988. "Robert Moses: The Expressway World." In *All That is Solid Melts Into Air: The Experience of Modernity*. New York: Viking Penguin.

Berry, Glen A. 1992. "House Music's Development and the East-Coast Underground Scene." MA Thesis, Madison: University of Wisconsin Press.

Bhatt, Sanjay. 2007. "How Other Cities Handle Clash of Clubs and Condos." *Seattle Times*, August 7.

Bianchini, Franco. 1995. "Night Cultures, Night Economies." *Planning Practice and Research* 10 (2): 121–26.

Blomley, Nicholas K. 1994. *Law, Space, and the Geographies of Power*. New York: Guilford Press.

———. 2003. Law, Property, and the Geography of Violence: The Frontier, the Survey, and the Grid. *Annals of the Association of American Geographers* 93 (1): 121–41.

———. 2004. *Unsettling the City: Urban Land and the Politics of Property*. New York: Routledge.

Boddy, Trevor. 1992. "Underground and Overhead: Building the Analogous City." In *Variations on a Theme Park: The New American City and the End of Public Space*, edited by M. Sorkin, 123–53. New York: Hill and Wang.

Bookman, Robert. 1989. "New York's Cabarets Don't Deserve Reputation for Violence." *New York Times*, June 3.

Böse, M. 2003. "Manchester's Cultural Industries: A Vehicle of Racial Ex/inclusion?" In *Consumption and the Post-Industrial City*, edited by F. Feckardt and D. Hassenpflug, 167–87. Frankfurt: Peter Lang Publishers.

Bourdieu, Pierre. 1984. *Distinction: Social Critique of the Judgement of Taste.* Cambridge, MA: Harvard University Press.

Boyd, Christopher M.J. 2009. "Can a Marxist Believe in Human Rights?" *Critique* 37 (4): 579–600.

Brash, Julian. 2011. *Bloomberg's New York: Class and Governance in the Luxury City.* Athens: University of Georgia Press.

Brenner, Neil, Jamie Peck, and Nik Theodore. 2010. "Variegated Neoliberalization: Geographies, Modalities, Pathways." *Global Networks* 10 (2): 182–222.

Brenner, Neil, and Nik Theodore. 2002. "Cities and the Geographies of 'Actually Existing Neoliberalism.'" *Antipode* 34 (3): 349–79.

Brown, Ethan. 2004. "Closing Time?", *New York Magazine*, February 23.

Brown, Wendy. 2003. "Neo-liberalism and the End of Liberal Democracy." *Theory & Event* 7 (1): 1–29.

Buckland, Fiona. 2002. *Impossible Dance: Club Culture and Queer World-making.* Middletown, CT: Wesleyan University Press.

Bumiller, Elisabeth. 1996. "Can Clubland Live in Quality-of-Life Era?" *New York Times*, August 4.

Burger, Michael. 2002. "Noise Off." *Gotham Gazette*, November 18.

Campo, Daniel, and Brent D. Ryan. 2008. "The Entertainment Zone: Unplanned Nightlife and the Revitalization of the American Downtown." *Journal of Urban Design* 13 (3): 291–315.

Carlson, Jen. 2007. "Tonic Goes To City Hall." *Gothamist*, April 18.

Caulfield, Jon. 1989. "'Gentrification' and Desire." *Canadian Review of Sociology/Revue canadienne de sociologie* 26 (4): 617–32.

Chatterton, Paul, and Robert Hollands. 2003. *Urban Nightscapes: Youth Cultures, Pleasure Spaces and Corporate Power.* London: Routledge.

Cheren, Mel, and Gabriel Rotello. 2000. *My Life and the Paradise Garage: Keep on Dancin'.* New York: 24 Hours for Life.

Chevigny, Paul. 1991. *Gigs: Jazz and the Cabaret Laws in New York City.* New York: Routledge.

———. 2004. "Social Dancing and Social Association." Unpublished paper, Document on file with author.

Chouinard, Vera. 1994. "Geography, Law and Legal Struggles: Which Ways Ahead?" *Progress in Human Geography* 18 (4): 415–40.

Christopherson, Susan. 1994. "The Fortress City: Privatized Spaces, Consumer Citizenship." In *Post-Fordism: A Reader*, edited by A. Amin, 409–27. Oxford: Blackwell.

Clough, N.L., and R.M. Vanderbeck. 2006. "Managing Politics and Consumption in Business Improvement Districts: The Geographies of Political Activism on Burlington, Vermont's Church Street Marketplace." *Urban Studies* 43 (12): 2261–84.

Cole, David. 1999. Hanging With the Wrong Crowd: Of Gangs, Terrorists, and the Right of Association. *Supreme Court Review*: 203–52.

CPC (City Planning Commission). 1989. *New York City: City Planning Comission.* The City of New York.

Currid, Elizabeth. 2007. *The Warhol Economy: How Fashion, Art, and Music Drive New York City.* Princeton, NJ: Princeton University Press.

Davis, Mike. 1992. *City of Quartz: Excavating the Future in Los Angeles*. New York: Vintage Books.
DCA (Department of Consumer Affairs). 2003. "Why A New Nightlife License Law." The City of New York.
———. 2004. "Transcript of the 2003 DCA Public Hearing on Cabaret Regulations." The City of New York
DCP (Department of City Planning). 1990. "Zoning Handbook: A Guide to New York City's Zoning Resolution." City of New York.
Debord, Guy. 1994. *The Society of the Spectacle*. New York: Zone Books.
Deutsch, Claudia. 1992. "Eviction Suit Threatens Bitter End, a Folk Legend." *New York Times*, September 28.
Dikeç, Mustafa, and Liette Gilbert. 2002. "Right to the City: Homage or a New Societal Ethics?" *Capitalism, Nature, Socialism* 13 (2): 58–74.
Duane, Thomas K. 1993. "Save Our Discos." *New York Times*, July 3.
Duke, Joanna. 2009. "Mixed Income Housing Policy and Public Housing Residents 'Right to the City.'" *Critical Social Policy* 29 (1): 100.
Dunlap, David W. 1983. "Two-Thirds of City's Cabarets Violating the Fire Safety Law." *New York Times*, March 28.
———. 1988. "Clubs Shatter Peace of Gramercy Park." *New York Times*, June 20.
———. 1989. "Club Owners Balk at Zoning Proposal." *New York Times*, June 22.
Dyer, Richard. 1979. "In Defence of Disco." *Gay Left* 8: 20–23.
Editorial. 1986. "New York's Ill-Tuned Cabaret Law." *New York Times*. June 26.
Ehrenreich, Barbara. 2007a. "Dance, Dance, Revolution." *New York Times*, June 3.
———. 2007b. *Dancing in the Streets: A History of Collective Joy*. 1st ed. New York: Metropolitan Books.
Eldridge, Adam, and Marion Roberts. 2008. "A Comfortable Night Out? Alcohol, Drunkenness and Inclusive Town Centres." *Area* 40 (3): 365–74.
Ellickson, Robert C. 1996. "Controlling Chronic Misconduct in City Spaces: Of Panhandlers, Skid Rows, and Public-Space Zoning." *The Yale Law Journal* 105 (5): 1165–248.
Fainstein, Susan S. 1994. *The City Builders: Property, Politics, and Planning in London and New York*. Oxford, UK: Blackwell.
Fainstein, Susan S., and Robert James Stokes. 1998. "Spaces For Play: The Impacts of Entertainment Development on New York City." *Economic Development Quarterly* 12 (2): 150–65.
Farber, Jim. 1995. "There's Trouble in Clubland." *Daily News*, May 15.
Fawaz, Mona. 2009. "Neoliberal Urbanity and the Right to the City: A View from Beirut's Periphery." *Development and Change* 40 (5): 827–52.
Feagin, Joe R. 1998. "Arenas of Conflict: Zoning and Land-Use Reform in Critical Political-Economic Perspective." In *The New Urban Paradigm: Critical Perspectives on the City*. Lanham, MD: Rowman & Littlefield Publishers.
Fecht, John. 2004. "New York Mayor in Fight Against Noise Pollution." *City Mayors*, June 10, http://www.citymayors.com/environment/nyc_noise.html.
Fernandes, E. 2007. "Constructing the 'Right to the City' in Brazil." *Social & Legal Studies* 16 (2): 201–19.
Ferris, Marc. 2007. "Moms and Dads in West Chelsea Clubland." *The Real Deal*, November 1.
Festa Defendants, 2005a, "Defendants' Memorandum of Law in Support of Cross-Motion for Summary Judgment and in Opposition to Plaintiffs' Motion for a Preliminary Injunction." Department of Law, City of New York. September 1.
———. 2005b, "Defendants' Reply Memorandum of Law in Further Support of Cross-Motion for Summary Judgment." Department of Law, City of New York. December 8.

Festa Plaintiffs, 2005a, "Plaintiffs' Memorandum of Law in Support of Motion for Preliminary Injunction." Document on file with Paul Chevigny. Date missing.

———. 2005b, "Plaintiffs' Memorandum of Law in Opposition to Defendants' Cross-Motion for Summary Judgment and Reply Memorandum in Support of Plaintiffs' Motion for a Preliminary Injunction." Document on file with author. October 12.

Fikentscher, Kai. 2000. *"You Better Work!": Underground Dance Music in New York City.* Hanover, NH: University Press of New England.

Fischler, Raphael. 2001. "Toward a Genealogy of Planning: Zoning and the Welfare State. *Planning Perspectives* 13 (4): 389–410.

Fleetwood, Blake. 1979. "The New Elite and an Urban enaissance." *New York Times*, January 14.

Florida, Richard L. 2004. *The Rise of the Creative Class: And How It's Transforming Work, Leisure, Community and Everyday Life.* New York: Basic Books.

Forman, Seth. 2000. "Community Boards." *Gotham Gazette*, September 20.

Fowler, Glenn. 1976. "Planning Unit Asks SoHo-NoHo Discotheque Ban." *New York Times*, August 12.

———. 1979. "Restoration of Times Sq. Moving Closer to Reality." *New York Times*, June 18.

Freedman, Samuel. 1986. "Musicians Taking City Cabaret Law to Court." *New York Times*, June 4.

Freeman, Joshua Benjamin. 2000. *Working-Class New York: Life and Labor Since World War II.* New York: New Press: Distributed by W.W. Norton.

Freitag, Michael. 1989. "Neighbors Complain As Nightclub Crowds Turn to Violence." *New York Times*, May 7.

Freitag, Michael. 1990. "Violence at Discotheque Mobilizes Neighborhood." *New York Times*, January 8.

Fyfe, Nicholas R., and Jon Bannister. 1996. "City Watching: Closed Circuit Television Surveillance in Public Spaces." *Area* 28 (1): 37–46.

Giroux, Henry A. 2006. *Stormy Weather: Katrina and the Politics of Disposability.* Boulder, CO: Paradigm Publishers.

Glaberson, William. 1990. "Irritated Neighbors Are Pressuring Discos." *New York Times*, September 7.

Goldstein, Richard. 1997. "Whose Quality of Life Is It, Anyway?" *Village Voice*, February 18.

Goss, Jon. 1993. "The 'Magic of the Mall': An Analysis of Form, Function, and Meaning in the Contemporary Retail Built Environment." *Annals of the Association of American Geographers* 83 (1): 18–47.

Gotham, Kevin Fox. 2005. "Tourism Gentrification: The Case of New Orleans' Vieux Carre (French Quarter)." *Urban Studies* 42 (7): 1099–121.

Gotham Gazette. 2005 "Giuliani—Savior of the City, or Its Enemy?" *Gotham Gazette*, August 8.

Grazian, D. 2005. *Blue Chicago: The Search for Authenticity In Urban Blues Clubs.* Chicago: University of Chicago Press.

Green, Mark. 1993. "New York City Can Have Its Discos and Restrict the Noise Too." *New York Times*, July 17.

Greenberg, Miriam. 2008. *Branding New York: How a City in Crisis Was Sold to the World.* New York: Routledge.

Gregory, Derek. 1994. *Geographical Imaginations.* Cambridge, MA: Blackwell.

Gross, Michael. 1985. "The Party Seems to be Over for Lower Manhattan Clubs." *New York Times*, October 26.

Hackworth, Jason. 2001. "Inner-City Real Estate Investment, Gentrification, and Economic Recession in New York City." *Environment and Planning A* 33 (5): 863–80. London: Pion Ltd.

———. 2007. *The Neoliberal City: Governance, Ideology, and Development in American Urbanism.* Ithaca, NY: Cornell University Press.

Hackworth, Jason, and Neil Smith. 2001. "The Changing State of Gentrification." *Tijdschrift voor Economische en Sociale Geografie* 92 (4): 464–77.

Hadfield, Phil. 2006. *Bar Wars: Contesting the Night in Contemporary British Cities, Clarendon Studies in Criminology.* Oxford: Oxford University Press.

———. 2008. "From Threat to Promise: Nightclub 'Security', Governance and Consumer Elites." *British Journal of Criminology* 48: 429–47.

———. (ed.) 2009. Nightlife and Crime: Social Order and Governance in International Perspective. Oxford; New York: Oxford University Press.

Hae, Laam. 2011a. "Rights to Spaces for Social Dancing in New York City: A Question of Urban Rights." *Urban Geography* 32 (1): 129–142

———. 2011b. "Dilemmas of the Nightlife Fix: Post-industrialization and the Gentrification of Nightlife in New York City." *Urban Studies*, 48 (15): 3444—3460.

Hall, Tim, and Phil Hubbard. 1998. *The Entrepreneurial City: Geographies of Politics, Regime, and Representation.* Chichester, UK: Wiley.

Hanna, Judith Lynne. 1979. *To Dance is Human: A Theory of Nonverbal Communication.* Austin: University of Texas Press.

Harvey, David. 1982. *The Limits to Capital.* Chicago: University of Chicago Press.

———. 1988."Voodoo cities." *New Statesman and Society* 1 (17):33-35.September 30.

———. 1989. "From Managerialism to Entrepreneurialism: The Transformation in Urban Governance in Late Capitalism." *Geografiska Annaler. Series B, Human Geography* 71 (1): 3–17.

———. 2000. "Uneven Geographical Developments and Universal Rights." In *Spaces of Hope.* Berkeley: University of California Press.

———. 2003a. "The Art of Rent: Globalization, Monopoly and the Commodification of Culture. In *Socialist Register 2002: A World of Contradictions*, edited by L. Panitch and C. Leys. Toronto: Fernwood Books Ltd.

———. 2003b. "The Right to the City." *International Journal of Urban and Regional Research* 27 (4): 939–41.

———. 2005. *A Brief History of Neoliberalism.* Oxford: Oxford University Press.

———. 2008. "The Right to the City." *New Left Review* 53, *http://www.newleftreview.org/?view=2740.*

Hatch, Trey. 2002. "Keep on Rockin'in the Free World: A First Amendment Analysis of Entertainment Permit Schemes." *Columbia Journal of Law & the Arts* 26: 313.

Hawkins, David. 1988. "If You're Thinking of Living in Flatiron District." *New York Times*, October 30.

Healy, Patrick. 2003. "Ways to Flout Cabaret Law As Varied As Dance Steps." *New York Times*, June 28.

Heath, Tim. 1997. "The Twenty-Four Hour City Concept—A Review of Initiatives in British Cities." *Journal of Urban Design* 2 (2): 193–204.

Henkin, Louis. 1968. "Foreword: On Drawing Lines." *Harvard Law Review* 82: 63–83.

Hennessy, Gamal. 2010. "Victims of Their Own Success: Nightlife as the Unwitting Agent of Gentrification." *New York Nights*, June 10.

Hill, Richard Child. 1984. "Fiscal Crisis, Austerity Politics, and Alternative Urban Policies." In *Marxism and the Metropolis*, edited by W. K. Tabb and L. Sawers, 298–322. New York: Oxford University Press.

Hobbs, Dick, Simon Winlow, Philip Hadfield, and Stuart Lister. 2005. "Violent Hypocrisy." *European Journal of Criminology* 2 (2): 161–83.

Horsley, Carter B. 1978. "The Impact of Disco Mania on Nightspots Is Growing." *New York Times*, July 9

Horsley, Carter B. 1979. "Zoning Controls on Discos Planned: Zoning Controls Being Planned to Cover Discos in the City." *New York Times*, November 4.

Howe, Marvine. 1992a "Chelsea Journal; Discomania? Some Say It's Really Discoterror." *New York Times*, June 4.

———. 1992b. "Cabaret Law Would Aid Neighbours of Nightclubs." *New York Times*, July 26.

———. 1993. "Noisy Nightclubs and Their Neighbors." *New York Times*, September 26.

Hu, Winnie. 2005. "City Noise Code Gets Stricter; Fancy Meter Will Aid Ears." *New York Times*, December 22.

———. 2006. "Clubgoers Fear Losing Privacy In Quinn Plan For Cameras." *New York Times*, August 14.

Hubbard, Phil. 2004. "Cleansing the Metropolis: Sex Work and the Politics of Zero Tolerance." *Urban Studies* 41 (9): 1687–1702.

Huden, Johanna. 2002. "The City Vs. Dancing." *New York Post*, November 13.

Huq, Rupa. 1999. "The Right to Rave: Opposition to the Criminal Justice and Public Order Act 1994." In *Storming the Millennium: The New Politics of Change*, edited by T. Jordan and A. Lent, 15–33. London: Lawrence & Wishart.

Hutson, Scott R. 2000. "The Rave: Spiritual Healing in Modern Western Subcultures." *Anthropological Quarterly* 73 (1): 35–49.

Jackson, Kenneth. 2005. "The Second Most Effective Mayor of the 20th Century." *Gotham Gazette*.

Jayne, Mark, Sarah L. Holloway, and Gill Valentine. 2006. "Drunk and Disorderly: Alcohol, Urban Life and Public Space." *Progress in Human Geography* 30 (4): 451–68.

Jayne, Mark, Gill Valentine, and Sarah L Holloway. 2008. "Geographies of Alcohol, Drinking and Drunkenness: A Review of Progress." *Progress in Human Geography* 32 (2): 247–63.

———. 2010. "Emotional, Embodied and Affective Geographies of Alcohol, Drinking and Drunkenness." *Transactions of the Institute of British Geographers* 35 (4): 540–54.

Jessop, Bob. 2002. *The Future of the Capitalist State*. Cambridge, UK: Polity.

Johnson, Richard. 2002. "Page Six: Bloomy Blasted on Club Law." *New York Post*, October 27.

Johnson, Wesley. 2010. "Third of Under-24s 'Drink to Get Drunk.'" *Independent*, September 7.

Jordan, J. 1998. "The Art of Necessity: The Subversive Imagination of Anti-Road Protest and Reclaim the Streets." In *DiY Culture: Party & Protest in Nineties Britain*, edited by G. McKay, 129–51. London: Verso.

Judd, Dennis R. 1995. "The Rise of the New Walled Cities." In *Spatial Practices: Critical Explorations in Social/Spatial Theory*, edited by H. Liggett and D. C. Perr, 144–68. Thousand Oaks, CA: Sage.

Katz, Lina. 2000. "Weekend Mambo March Aims to Bump and Grind Cabaret Law into the Ground." *Village Voice*, August 29.

Kearns, Gerard, and Chris Philo. 1993. *Selling Places: The City as Cultural Capital, Past and Present*. Oxford: Pergamon Press.

Kennedy, Shawn G. 1976. "The New Discotheque Scene: 'Like Going to a Big House Party.'" *New York Times*, January 3.

———. 1983. "New Move to Reduce City Noise." *New York Times*, April 30.

Klein, Naomi. 2000. *No Logo: Taking Aim at the Brand Bullies*. Toronto: Knopf Canada.

Koch, Edward. 1988. "Communiqué." Office of the Mayor. The City of New York.

Kohn, Margaret. 2004. *Brave New Neighborhoods: The Privatization of Public Space*. New York: Routledge.
Landis, Dylan. 1983. "If you're thinking of living in TriBeCa." *New York Times*, May 15.
Larner, Wendy. 2000. "Neo-liberalism: Policy, Ideology, Governmentality." *Studies in Political Economy* 63: 5–26.
Lawrence, Tim. 2003. *Love Saves the Day: A History of American Dance Music Culture, 1970–1979*. Durham, NC: Duke University Press.
Lawrence, Tim. 2006. "In Defense of Disco (Again)." *New Formations* 58: 128–46.
LeDuff, Charlie. 1996. "The New beat? Take Out Da Funk, Take Out Da Fun." *New York Times*, June 2.
Lees, Loretta. 2008. "Gentrification and Social Mixing: Towards an Inclusive Urban Renaissance?" *Urban Studies* 45 (12): 2449–2470.
Lees, Loretta, Tom Slater, and Elvin K. Wyly. 2007. *Gentrification*. New York: Routledge.
———. 2010. *The Gentrification Reader*. London: Routledge.
Lefebvre, Henri. 1991. *The Production of Space*. Oxford: Blackwell.
———. 1996. "The Right to the City." In *Writings on Cities*, edited by E. LeBas and E. Kofman, 63–181. Cambridge, MA: Blackwell.
———. 2003. *The Urban Revolution*: Minneapolis: University of Minnesota Press.
Levy, David C. 1986. "New York Mistreats Jazz." *New York Times*, April 12.
Ley, David. 1996. *The New Middle Class and the Remaking of the Central City*. Oxford: Oxford University Press.
Lichten, Eric. 1986. *Class, Power, & Austerity: The New York City Fiscal Crisis*. South Hadley, MA: Bergin & Garvey Publishers.
Lipietz, Alan. 1994. "Post-Fordism and Democracy." In *Post-Fordism: A Reader*, edited by A. Amin, 338–57. Oxford: Blackwell.
Lloyd, Richard D. 2006. *Neo-Bohemia: Art and Commerce in the Postindustrial City*. New York: Routledge.
Lovatt, Andy, and Justin O'Connor. 1995. "Cities and the Night Time Economy." *Planning Practice and Research* 10 (2): 127–34.
Lukes, Steven. 1985. *Marxism and Morality*. Oxford: Clarendon Press.
———. 1991. *Moral Conflict and Politics*. Oxford: Clarendon Press.
MacLeod, Gordon. 2002. "From Urban Entrepreneurialism to a 'Revanchist City'? On the Spatial Injustices of Glasgow's Renaissance." *Antipode* 34 (3): 620–24.
Malbon, Ben. 1999. *Clubbing: Dancing, Ecstasy and Vitality*. London: Routledge.
Marcuse, Peter. 2004. "The 'War on Terrorism' and Life in Cities after September 11, 2001." In *Cities, War, and Terrorism: Towards an Urban Geopolitics*, edited by S. Graham, 263–75. Malden, MA: Blackwell Publishing.
Mayer, Margit. 2009. "The 'Right to the City' in the Context of Shifting Mottos of Urban Social Movements. *City* 13 (2): 362–74.
McArdle, Andrea, and Tanya Erzen. 2001. *Zero Tolerance: Quality of Life and the New Police Brutality in New York City*. New York: New York University Press.
McHugh, Clare. 1989. "For Night Crawlers, Are City's Excitement Fading? The Big-Name Entertainers Now Shun the NY Clubs; Big Discos Dead, Too." *New York Observer*, December 11.
McKay, George. 1998. *DiY Culture: Party & Protest in Nineties Britain*. London: Verso.
McNally, David. 2001. *Bodies of Meaning: Studies on Language, Labor, and Liberation*. Albany, NY: State University of New York Press.
Mele, Christopher. 2000. *Selling the Lower East Side: Culture, Real Estate, and Resistance in New York City, Globalization and Community*. Minneapolis: University of Minnesota Press.

Merrifield, Andy. 2002. *Dialectical Urbanism: Social Struggles in the Capitalist City*. New York: Monthly Review Press.
Mitchell, Don. 2003. *The Right to the City: Social Justice and the Fight for Public Space*. New York: Guilford Press.
———. 2005. "Property Rights, the First Amendment, and Judicial Anti-Urbanism: The Strange Case of Virginia v. Hicks." *Urban Geography* 26 (7): 565–86.
Mitchell, Don, and Nik Heynen. 2009. "The Geography of Survival and the Right to the City: Speculations on Surveillance, Legal Innovation, and the Criminalization of Intervention." *Urban Geography* 30 (6): 611–32.
Mitchell, Katharyne, and Katherine Beckett. 2008. "Securing the Global City: Crime, Consulting, Risk, and Ratings in the Production of Urban Space." *Indiana Journal of Global Legal Studies* 15 (1): 75–100.
Mollenkopf, Jon H., and Manuel Castells, eds. 1991. *Dual City: Restructuring New York*. New York: Russell Sage Foundation Publications.
Moynihan, Colin. 1999. "Neighborhood Report: Lower East Side; In Defense of the Inalienable Right to . . . Boogie." *New York Times*, January 10.
———. 2009. "Punk Institution Receives City Money for New Building." *New York Times*, June 28.
Musto, Michael. 1987. "The Death of Downtown." *Village Voice*, April 28.
Narvaez, Alfonso. 1971. "City Acts to Let Homosexuals Meet and Work in Cabarets." *New York Times*, October 12.
———. 1976. "When Bars Close, Night is Young at Illegitimate but Abundant Clubs." *New York Times*, December 26.
Neuberg, Eva. 2001. "No Dancing Allowed: Protesting the City's Cabaret Laws Outlawing Nightlife." *New York Press*, August 22–28.
Newfield, Jack. 2002. *The Full Rudy: The Man, the Myth, the Mania*. New York: Thunder's Mouth Press/Nation Books.
Newman, Kathe, and Elvin K. Wyly. 2006. "The Right to Stay Put, Revisited: Gentrification and Resistance to Displacement in New York City." *Urban Studies* 43 (1): 23–57.
Newman, Oscar. 1972. *Defensible Space; Crime Prevention Through Urban Design*. New York: Macmillan.
Niemietz, Brian. 2006. "Rich Man, Pour Man." *New York Post*, March 22.
Nieves, Evelyn. 1990. "Noisy Discos Are Targets of Crackdown." *New York Times*, August 20.
Noble, Presley B. 1989. "If You're Thinking of Living in TriBeCa." *New York Times*, May 14.
NYNA (New York Nightlife Association). 2004. "The $9 Billion Economic Impact of the Nightlife Industry on New York City: A Study of Spending By Bar/Lounges and Clubs/Music Venues and Their Attendees." New York: New York Nightlife Association.
O'Brien, Elizabeth. 2003. "Looking to End Tap Dance Around Cabaret Laws." *Downtown Express*, July 8–14.
Ocejo, Richard E. 2009. "City Nights: The Political Economy of Postindustrial Urban Nightlife." PhD diss., City University of New York.
Oelsner, Lesley. 1978. "'New' Times Square Waiting in the Wings." *New York Times*, November 14.
Office of Councilor Gerson. 2007. "Tonic Bill Meeting Minutes." New York City. Document on file with author.
Office of the Mayor, City of New York. 2002. "Mayor Michael R. Bloomberg Announces Operation Silent Night: Quality of Life Initiative Targets Areas Plagued By Loud and Excessive Noise," news release, October 2, home2.nyc.gov/html/om/html/2002b/pr257-02.html.

Ong, Aihwa. 2007. "Neoliberalism As a Mobile Technology." *Transactions of the Institute of British Geographers* 32 (1): 3–8.

Owen, Frank. 1997a. "Crackdown in Clubland: City Hall Is Changing the Rules of Nightlife in New York." *Village Voice*, February 18.

———. 1997b. "Dry Zone." *Village Voice*, June 17.

Palmer, Bryan D. 2000. *Cultures of Darkness: Night Travels in the Histories of Transgression*. New York: Monthly Review Press.

Papayanis, Marilyn Adler. 2000. "Sex and the Revanchist City: Zoning Out Pornography in New York." *Environment and Planning D: Society and Space* 18: 341–53.

Pareles, Jon. 1986. "City Forces Clubs to Restrict Jazz." *New York Times*, April 10.

———. 1987. "At Unlicensed Jazz Clubs, 3 is a Crowd but 4 Is Illegal." *New York Times*, March 19.

Peck, Jamie, and Adam Tickell. 1994. "Searching for a New Institutional Fix: The After-Fordist Crisis and the Global-Local Disorder." In *Post-Fordism: A Reader*, edited by A. Amin, 280–315. Oxford: Blackwell.

Peck, Jamie, and Adam Tickell. 2002. "Neoliberalizing Space." *Antipode* 34 (3):380–404.

Ponzini, Davide, and Ugo Rossi. 2010. "Becoming a Creative City: The Entrepreneurial Mayor, Network Politics and the Promise of an Urban Renaissance." *Urban Studies* 47 (5): 1037.

Pratt, Geraldine. 2005. "Abandoned Women and the Spaces of the Exception." *Antipode* 37 (5): 1053–78.

Purcell, Mark. 2003. "Citizenship and the Right to the Global City: Reimagining the Capitalist World Order." *International Journal of Urban and Regional Research* 27 (3): 564–90.

Raco, Mike, and Rob Imrie. 2000. "Governmentality and Rights and Responsibilities in Urban Policy." *Environment and Planning A* 32 (12): 2187–204.

Ramos, Josell. 2005. *Maestro*. United States: ARTrution. Sanctuary Records Group. DVD.

Ranasinghe, Prashan, and Mariana Valverde. 2006. "Governing Homelessness through Land-Use: A Sociolegal Study of the Toronto Shelter Zoning By-Law." *The Canadian Journal of Sociology/Cahiers canadiens de sociologie* 31 (3): 325–49.

Redhead, Steve. 1993. *Rave Off: Politics and Deviance in Contemporary Youth Culture*. Aldershot, UK: Avebury.

Redhead, Steve, Derek Wynne, and Justin O'Connor. 1998. *The Clubcultures Reader: Readings in Popular Cultural Studies*. Oxford: Blackwell.

Reed, Susan A. 1998. The Politics and Poetics of Dance. *Annual Review of Anthropology* 27: 503–32.

Reichl, Alexander J. 1999. *Reconstructing Times Square: Politics and Culture in Urban Development*, *Studies in Government and Public Policy*. Lawrence: University Press of Kansas.

Rhode, David. 1998. "Vowing Noise Crackdown, City Shuts Four East Side Bars." *New York Times*, April 7.

Roberts, Marion. 2006. "From 'Creative City' to 'No-Go Areas': The Expansion of the Night-Time Economy in British Town and City Centres." *Cities* 23 (5): 331–38.

Romano, Tricia. 2002. "The Safety Dance: You Can't Dance If You Want to." *Village Voice*, November 27.

———. 2003a. "Fly life: Dance! Dance! Dance! (Yowza! Yowza! Yowza!): Say Good-Bye to the Cabaret Law." *Village Voice*, November 24.

———. 2003b. "Fly Life: Who Will Become the New Dance Police?" *Vllage Voice*, December 3.

———. 2004. "The Next Brooklyns." *Village Voice*, July 22.

Rose, Damaris. 1984. "Rethinking Gentrification-Beyond the Uneven Development of Marxist Urban Theory." *Environment and Planning D: Society Space* 2 (1): 47–74.

Rose, Nikolas. 1996. "The Death of the Social? Re-figuring the Territory of Government." *Economy and society* 25 (3): 327–56.
Rose, Nikolas, and Mariana Valverde. 1998. "Governed by Law?" *Social & Legal Studies* 7 (4): 541–51.
Rothman, Robin. 1999. "Sites Unscene." *Village Voice*, September 28.
Salkin, Allen. 2007. "Lower East Side Is under a Groove." *New York Times*, June 3.
Scheingold, Stuart A. 1974. *The Politics of Rights: Lawyers, Public Policy, and Political Change*. New Haven: Yale University Press.
Schuman, Wendy. 1974. "SoHo a 'Victim of its Own Success.'" *New York Times*, November 24.
Sharma, Nandita. 2010. "On Being Not Canadian: The Social Organization of 'Migrant Workers' in Canada." In *Colonialism and Racism in Canada: Historical Traces and Contemporary Issues*, edited by M. A. Wallis, Sunseri, L. and Galabuzi, G. Toronto: Nelson Education Ltd.
Shaw, Robert. 2010. "Neoliberal Subjectivities and the Development of the Night Time Economy in British Cities." *Geography Compass* 4 (7): 893–903.
Silverman, Jonathan. 2007. "City's Fading Club Scene." *Newsday*, May 4.
Sleeper, Jim. 1987. "Days of the Developers, Boom and Bust with Ed Koch." *Dissent* Fall: 437–50.
Smith, Neil. 1979. "Toward a Theory of Gentrification a Back to the City Movement by Capital, Not People." *Journal of the American Planning Association* 45 (4): 538–48.
———. 1982. "Gentrification and Uneven Development." *Economic geography* 58 (2): 139–55.
———. 1996. *The New Urban Frontier: Gentrification and the Revanchist City*. London: Routledge.
———. 1998. "Giuliani Time: The Revanchist 1990s." *Social Text* 16 (4): 1–20.
Snyder, Cara Renee. 2005. "Waking Up in the 'City that Never Sleeps': A Solution for Nighttime Noise in West Chelsea." *Columbia Journal of Environmental Law* 30 (1): 249–91.
Solnit, Rebecca, and Susan Schwartzenberg. 2000. *Hollow City: The Siege of San Francisco and the Crisis of American Urbanism*. London: Verso.
Sorkin, Michael. 1992. *Variations on a Theme Park: The New American City and the End of Public Space*. New York: Hill and Wang.
Sorkin, Michael, and Sharon Zukin. 2002. *After the World Trade Center: Rethinking New York City*. New York: Routledge.
Span, Paula. 1998. "On a Cleaning Spree: Nightclubs the New Target." *Washington Post*, February 20.
Spencer, Paul. 1985. *Society and the Dance: The Social Anthropology of Process and Performance*. Cambridge, UK: Cambridge University Press.
Staeheli, Lynn A., and Don Mitchell. 2006. "USA's Destiny? Regulating Space and Creating Community in American Shopping Malls." *Urban Studies* 43 (5–6): 977–92.
Steinhauer, Jennifer. 2002. "City Cracks Down on Nightclubs and May Revise Its Policies." *New York Times*, November 10.
———. 2004. "Bar Owners Say Mayor Wants Tucked-In City." *New York Times*, February 7.
Strom, Stephanie. 1990. "Hispanic Residents Rally Against Closing of Social Clubs." *New York Times*, April 6.
Swyngedouw, Eric. 1992. "The Mammon Quest: 'Glocalisation', Interspatial Competition and the Monetary Order: The Construction of New Scales." In *Cities and Regions in the New Europe*, edited by M. Dunford and G. Kakalas, 39–68. London: Belhaven Press.

Tabb, William K. 1982. *The Long Default: New York City and the Urban Fiscal Crisis*. New York: Monthly Review Press.

———. 1984. "The New York City Fiscal Crisis." In *Marxism and the Metropolis: New Perspectives in Urban Political Economy*, edited by W. K. Tabb and L. Sawers, 2nd ed, 323–45. New York: Oxford University Press.

Talbot, Deborah. 2004. "Regulation and Racial Differentiation in the Construction of Night-time Economies: A London Case Study." *Urban Studies* 41 (4): 887–901.

———. 2006. "The Licensing Act 2003 and the Problematization of the Night-Time Economy: Planning, Licensing and Subcultural Closure in the UK." *International Journal of Urban and Regional Research* 30 (1): 159–71.

Tantum, Bruce. 2004. "The Beat Goes On." *Time Out New York*, February 12.

Thomas Jr., Robert McG. 1974. "Weekly Parties for 500 Chill Tenants." *New York Times*, May 21.

Thompson, E. P. 1975. *Whigs and Hunters: The Origin of the Black Act*. 1st American ed. New York: Pantheon Books.

Thornton, Sarah. 1996. *Club Cultures: Music, Media, and Subcultural Capital*. 1st U.S. ed. Hanover, NH: University Press of New England.

Tretter, Eliot M. 2009. "The Cultures of Capitalism: Glasgow and the Monopoly of Culture." *Antipode* 41 (1): 111–32.

Tushnet, Mark. 1984. "An Essay on Rights." *Texas Law Review* 62: 1363–1412.

Wacquant, Loïc. 2001. "Deadly Symbiosis: When Ghetto and Prison Meet and Mesh." *Punishment and Society* 3 (1): 95–134.

Ward, Kevin. 2007. "'Creating a Personality for Downtown': Business Improvement Districts in Milwaukee." *Urban Geography* 28 (8): 781–808.

Ward, Stephen V. 1998. *Selling Places: The Marketing and Promotion of Towns and Cities, 1850–2000*. London: E & FN Spon/New York: Routledge.

Weber, Rachel. 2002. "Extracting Value from the City: Neoliberalism and Urban Redevelopment." *Antipode* 34 (3): 519–40.

Weller, Sheila. 1975. "New Wave of Discotheques." *New York Sunday News*, August 31.

Wilson, James Q., and George L. Kelling. 1982. "Broken Windows." *Atlantic Monthly* 249 (3): 29–38.

World Beat. 2001, "NYC's Legendary Wetlands Preserve Rock Club Forced to Close its Doors After Almost 13 Years," *New York Rock*, July 30.

Yarrow, Andrew L. 1989. "New Zoning Strictures Proposed for Nightclubs." *New York Times*, April 15.

Young, Iris Marion. 1990. *Justice and the Politics of Difference*. Princeton, NJ: Princeton University Press.

Zukin, Sharon. 1989. *Loft Living: Culture and Capital in Urban Change*. Baltimore: Johns Hopkins University Press.

———. 1991. *Landscapes of Power: From Detroit to Disney World*. Berkeley: University of California Press.

———. 1995. *The Cultures of Cities*. Cambridge, MA: Blackwell.

———. 2009. *Naked City: The Death and Life of Authentic Urban Places*: New York: Oxford University Press.

Cases Cited

Barnes v. Glen Theater, Inc. 1991, 501 US 560. No. 90–26
Chiasson v. New York City Dept. of Consumer Affairs, 1986, 505 N.Y.S. 2d 499 (N.Y. Sup. Ct.)
Chiasson v. New York City Dept. of Consumer Affairs, 1988, 524 N.Y.S. 2d 629 (N.Y. Sup. Ct.)
City of Chicago v Morales, 1999, 527 U.S. 41. No. 97–1121
City of Dallas v. Stanglin, 1989, 490 U.S. 19. No. 87–1848
Festa. v. New York City Department of Consumer Affairs, 2006, NY Slip Op 26125 [12 Misc 3d 466]
Hicks v. Commonwealth, 2000, 33 Va. App. 561.
Hicks v. Commonwealth, 2001, 36 Va. App. 49.
Hicks v. Commonwealth of Virginia, 2002, 264 Va. 48.
Virginia v. Hicks, 2003, 123 S. Ct. 2191

Index

For Product Safety Concerns and Information please contact our EU representative GPSR@taylorandfrancis.com Taylor & Francis Verlag GmbH, Kaufingerstraße 24, 80331 München, Germany